The Cults of Relativity

Finding Einstein, Twain and a Universe Beyond $E=mc^2$

Drake Larson
& Nora De Caprio

Hellgate Press
Ashland, Oregon

The Cults of Relativity

©2008 by Drake Larson
Published by L&R Publishing/Hellgate Press

Hellgate Press
An Imprint of L&R Publishing, LLC
P.O. Box 3531
Ashland, Oregon 97520
1-800-795-4059
www.hellgatepress.com

Book designed & edited by Harley B. Patrick
Cover design by Fred Daniel

Library of Congress Cataloging-in-Publication Data

      Larson, Drake, 1952-
  The cults of relativity : finding Einstein, Twain and a universe beyond E=mc2 / Drake Larson & Nora De Caprio.
    p. cm.
  Includes index.
  ISBN-13: 978-1-55571-651-6
  1. Relativity (Physics)--Popular works. I. DeCaprio, Nora. II. Title.
  QC173.57.L37 2008
  530.11--dc22

                2007039603

Printed and bound in the United States of America
First Edition 10 9 8 7 6 5 4 3 2 1

Contents

"Do not go where the path may lead, go instead where there is no path and leave a trail." - *Ralph W. Emerson (1803-1882)*

Preface

This book will explore quandaries tackled by Albert Einstein and Mark Twain. We hope this tale of how our heroes blazed new trails will entice others to do some trailblazing of their own. In this spirit, we'll depart from what most readers of popular science books have come to expect. Think of it as a literary throwback into the Amazon with Teddy Roosevelt, instead of a modern nature-walk or presidential tour.

Today's carefully choreographed presidential tours are essentially photo-ops. The president touring a natural disaster likely starts with a concerned wave to the cameras before hopping into a helicopter with official experts who brief him on the problems and solutions at hand. Everyone wears shiny hard hats and no one gets their hands dirty. The president's tour loops comfortably back to familiar cameras so he can assure everyone that our country's "best" are busy making the world safer for all of us. Then it's off with the shiny hard hat and into a sparkling tuxedo – that evening's photo-op – another lavish fundraiser.

Teddy Roosevelt preferred unpredictable expeditions to comfortable VIP tours. Our president who established the National Park Service would have also shunned today's sanitized trail loops in our national parks. These modern trails have conveniently excluded all of nature's dangers and merely display tranquil wildness. They have the ambiance of modern science books and VIP tours. Teddy Roosevelt got down to the nitty-gritty of true exploration – and that's exactly what this book will do.

Teddy's Amazon expedition had unfortunate members consumed alive by wildlife or drowned in swift currents. Deadly fever almost killed Teddy as he navigated through this steamy wilderness. One member of the expedition was so overwhelmed by all the unexpected "excitement" that he lost his sanity. He ran deep into the jungle never to be seen again...hmm...unpredictable expeditions are not for everyone.

So here's what you can expect on our literary expedition: Early chapters are rather comfortable tours about statistics and Relativity. Twain and Einstein are our guides into paradox. But be prepared for this thoughtful stroll to morph into an untamed expedition in Chapter 4 and beyond. Today's experts have warned us to keep out of these scientific swamps and we will be in true wilderness – abandoning a pleasant science tour and going where we're told not to go. In other words, expect to feel a little lost as unanswered questions overwhelm you. That said, why these mysteries of science exist is remarkably easy to appreciate.

As we explore this uncharted territory, we are sure to encounter a menacing beast to many casual readers of science – the equation. In truth, these mathematical creations are beautiful for many but beastly for those who do not understand that slow reading is needed to reveal their beauty. Equations are like Monet's varied dabs forming beautiful art. Granted, with a quick glance one immediately knows the colors on the canvas. But daydreaming transforms dabs into beauty while rushing denies insight and enjoyment. The same is true for eloquent equations. One has to enter a fantasy world to savor the numerical nuances and beauty of equations.

This ability to fantasize mathematically is invaluable when exploring new concepts. There exists no better example of daydreaming disjointed mysteries into eloquent equations than how Einstein discovered Relativity. And Einstein is not alone. Archimedes' famous solution came to him while he was taking a warm bath. Many great mathematicians mention how their famous discovery came to be while their mind was just drifting – sensing a pleasant answer from a twilight dance with intrigue.

So you have two options for the few equations in this book. One is to breeze through them and concentrate on the words that also convey the gist of our story, or change your pace of reading...linger and enjoy playful daydreams with equations.

Finally, we hope this book will encourage the mathematically inclined not to be too bashful in following their creative instincts. Einstein and Twain were both creative explorers who needed only pen and paper to craft their masterpieces. Feel free to get a pen and paper and have at it if one of our trails ends in a corner of science wilderness that sparks your interest. We'll guarantee you won't drown, catch fever or feel the need to run deep into the Amazon jungle. But mental exhaustion from intrigue... that's a definite possibility.

Introduction

Mark Twain's pioneering boldness in *The Adventures of Huckleberry Finn* is often called the birth of American literature, while Albert Einstein's boldness in physics was the birth of Relativity. Their similarities go well beyond being ordained as founding fathers. Each possessed the self-confidence to accomplish what others would consider too radical to even attempt. They both castigated statistics – were crusaders for social fairness – oblivious to unkempt hair and clutter as they dabbled in the other's arena of fame. Einstein took great pleasure in crafting aphorisms like, "If a messy desk is a sign of a messy mind, then what's an empty desk a sign of?" In kind – Twain tried his hand at deriving the true sizes of atoms and loved exploring the mysteries of electricity with his friend Nicolas Tesla.

One of Twain's last works warned of future atomic bombs decades before Einstein wrote his famous letter warning FDR that such weapons were possible.

Striking – two so similar were famed by so different as math and prose.

Einstein successfully challenged the conventional wisdom in physics – remaining true to his inspirations and equations even though they led him to seemingly absurd conclusions that the leading men in his field initially dismissed with condescension. Less well known is how Twain proved equally un-intimidated by math-derived conventional wisdom. At a time when just about all whites accepted the racist notion that blacks were genetically inferior, Twain wrote a daring book that made a hero of a black slave and openly lampooned prevailing elitist views. Irreverent courage lay behind Twain's dead-on wicked wisdom about those spewing statistics – encapsulated in his quip, "Figures don't lie, liars figure."

Both Einstein and Twain understood math's insights and limitations in ways that seem forgotten in today's statistics-driven world. We will explore this lost talent -- and find -- far more uncertainty and contradiction in contemporary statistics and physics than most experts care to admit.

We'll then take dead aim at wisdom-induced trances.

Twain

The cunning wit of Huck Finn and the slave Jim in **The Adventures of Huckleberry Finn** is great entertainment. This folksy triumph of two uneducated lower-class misfits also delivered a powerful attack on an intellectual fad then infecting America's prestigious universities.

Darwin's theory of evolution had been hotly debated for 25 years when Twain's masterpiece was published. Elitists were using Darwin's theory to create a new science justifying many of the inequities within society. These elitists claimed that the impersonal hand of natural selection placed those of superior intelligence into positions of inherited wealth and power while those born into poverty were simply social evolution's dimwits.

This science, called eugenics, also promised that intelligence could be calculated by the size and shape of the head. Lead researchers were so fortunate that their own cranial dimensions were available to set the benchmark for true genius. Complex statistical analysis of cranial profiling would prove to the public what was obvious to eugenics' discoverers. There were difficulties, like explaining the large cranial capacity of so many Afro-Americans. Researchers found correlating canine and water buffalo cranial dimensions most useful in solving this particular problem. Such nonsense justifying this theory used clever statistical analysis to transform randomness into a promising theory...and abracadabra!...nonsense masqueraded as brilliance.

Those teaching eugenics at institutions like Harvard were in an enviable power position. Bright young minds competed for approval by neatly regurgitating how complex cranial profiling will determine one's intelligence. Misguided professors teaching eugenics offered little more than a brainwashing for good students while bad grades admonished the less respectful ones. Then as now, this pretentious setting fostered a cult-like validation for Harvard's statistical wisdom.

The Adventures of Huckleberry Finn was a successful attack on elitist statistics that no mathematical counter argument could have accomplished. Twain would have gotten nowhere arguing in this huge maze of mumbo jumbo statistics from **the** most prestigious university in America...well...maybe as he was mired in math gobbledy-gook, he'd been put in his proper place with a smile of regret for not understanding the most brilliant elite.

But classic irony triumphed as prestigious condescension lost out to Huck, Jim, and the verb, ain't.

"There are three kinds of lies: Lies, Dammed Lies and Statistics," was Twain's succinct conclusion for elitist statistics – and the dangers of statistics in general. Dammed statistics pop up all the time and in unlikely places. An interesting modern example arose when the university that has the largest collection of Twain's original manuscripts set out to determine whether it had discriminated against women in graduate admissions. When totaled up university wide, the statistical analysis showed a clear pattern of discrimination against women. That seemed clear. So they carried the analysis further and checked the admission practices of each individual department. Lo and behold, *every* individual department in the university discriminated against men at the exact same time that the university, as a whole, was discriminating against women! The statisticians had used the same statistical method for each department as they'd used for the university as a whole! These impossible results may be perplexing for the university's governing counsel, but it's great for political debate. Politicians can pick the conclusion that suits their purpose. Opposite sides are armed with the following statistical facts: "Every department discriminated against men," versus "The University as a whole discriminates against women."

In Chapter 1, *Twain's World*, we'll look more closely at paradoxes in statistics as they relate to the world of medicine.

Einstein

In 1905 Einstein was the first to clearly explain some of the many mysteries in an uncharted territory now called Relativity. Einstein literally daydreamed his way to solving problems that had pestered some very bright and hard-working physicists for over a decade.

His solution was the warping of time and space.

Many other talented scientists quickly followed Einstein's head start into this new science of paradox and opportunity. All were hoping to be the first to discover a new phenomenon or explain some of the many mysteries of science. These frontiersmen were driven by enthusiastic competition reminiscent of the California Gold Rush. In 1916, Einstein discovered another rich mother lode of knowledge in Relativity by redefining a very useful tool in physics called, "The Conservation of Energy and Momentum" (Cem). Conservation of energy <u>and</u> momentum had become indispensable in electro-magnetic and classical physics. Cem concludes that energy is never gained or lost and momentum remains the same until otherwise acted upon. The beauty of Cem is how it enhances the exactness of the physics. It was only pragmatic to utilize this tool in Relativity.

Einstein's new concept of Cem was simple yet bold.

Meanwhile, other scientists were gaining knowledge in a new field called quantum mechanics. Everyday statistics and quantum mechanics are similar methods of using statistics to describe the probability of events in the real world around us. Relativity and quantum mechanics are different, semi-competing methods of describing and predicting the miniature world. Every good gambler knows that the flip-side of probability is uncertainty. Einstein questioned the wisdom of quantum mechanics relying so heavily on probability and statistics when so much of this new scientific wilderness of the 20th century had not yet been examined by more exact forms of mathematics. Einstein loathed abandoning the elegant, precise, and more prophetic mathematics that had served him so well for equations derived from statistics estimating the possibilities of random chance. Even today, many physicists don't understand the degree of uncertainty within probability and statistics that is so much of quantum mechanics. Mark Twain's, "Lies, Dammed Lies and Statistics" found an unexpected companion castigating statistics when Einstein quipped, "God does not play with dice."

Unfortunately, Einstein's more exact mathematical methods, called tensors, ran into serious difficulties before the early 1930s. Some experimental results fit his math predictions very well, while other experiments demonstrated flaws in these complicated equations. Worse yet, a few well-publicized "perfect" experimental results were later re-evaluated and retracted in a manner similar to the discovery of cold fusion.

Quantum mechanics has fared much better than have Einstein's tensor equations. Quantum's pragmatic mantra is, "find equations that fit the data and see how far they can be extended and remain relevant." Einstein's more ambitious goal – General Relativity becoming a prophetic unified theory that describes everything – has yet to materialize.

Today's physics students exposed to practical applications will use more electromagnetic and quantum physics than Relativity. General Relativity is taught almost as an aside at most universities because the 1873 equations of Maxwell describing electromagnetic physics also describe much of practical Relativity. Data-derived quantum equations often intertwine probability with Maxwell's physics. This subliminal exposure to the practical aspects of Relativity often precludes any serious study of its theoretical nuances, which are rich in paradoxes and tend to create more questions than answers. College professors often derive quantum equations by starting with their underlying foundation of probability and statistics. Then several incredible uses of these statistics describing chance are presented. It is almost irresistible not to mention the irony of how a genius' comment, "God does not play with dice," appears these days to be scientific heresy.

However, this slant of Einstein's quip lacks historical context. Relativity was in its infancy – a largely untouched territory – when Einstein's instinct was to utilize more exact mathematical methods than statistics. Our look at statistical paradoxes will show the basis for his reasoning. Today, statistics are used everywhere in the sciences, and have become an essential tool for mandating social justice. Most graduating college students don't appreciate the lurking danger of false conclusions that spring from "Lies, Dammed Lies and Statistics."

The Expedition's Map

Twain & Einstein's juncture is our portal to wilderness exploring as we wander into voids of knowledge quietly hidden in today's world.

Our expedition begins in Twain's world of statistics where paradoxes play havoc in the medical field. Twain loved both the capitalistic action and irony of trying to make the most bucks for the least effort. Our look at statistics in America's capitalistic, yet tightly regulated medical industry will give readers an appreciation for Twain's quips. His seemingly contradictory comments, "Figures don't lie, liars figure" and "Lies, Dammed Lies, and Statistics" will merge harmoniously.

Then we'll present a little history of physics to understand why Einstein made his bold revision to Cem and discovered $E=mc^2$. Our view of mathematical explanations to old questions will highlight why Twain and Einstein saw good reason to question so many truisms.

In Einstein's honor, we will take you step-by-step down the path he blazed to find his most famous discoveries. Einstein needed only the simplest tools of mathematics and his remarkable imagination. Our detailed description of his vision will be easily understandable for those comfortable with basic junior high school math skills.

But does Einstein always guide us in the right direction – or is there great opportunity to continue true wilderness exploring contrary to his underlying assumptions? Is some of physics' current conventional wisdom false? Phantom wisdom deceptively cloaked in complicated math analysis similar to the math mazes of eugenics?

We will ask other risky questions. Is the continual miniaturization of electronics leading us to a revolutionary new source of free energy – or are we encouraging a fruitless adaptation of the cold fusion crusade? Was Twain's fascination with annulling forces surrounding Tesla's cages a clue toward solving dilemmas in Einstein's physics? And more brazen still: Is a mind experiment the springboard for possible radical solutions to mysteries found in cosmology and quantum mechanics? What about one of its corollaries? Should we accept that the crucial second law of thermodynamics is not all encompassing, but more a yin to a mystery yang? We'll encourage bold physicists to follow their optimistic instincts.

The Power of Conventional Wisdom

We would like to remind the reader that resistance to new ideas contrary to conventional wisdom is remarkably powerful. Many great minds rejected Einstein's and Twain's new concepts – preferring to remain forever faithful to their antiquated conventional wisdom. This should be no great surprise. Galileo's contention that the earth revolves around the sun only convinced the intellectual elite of 1634 to initiate heresy charges. His reward for challenging the wisdom of his day was a lengthy house arrest. Galileo was actually quite lucky. A monk who made the same crazy claim years earlier was rewarded by being burned at the stake alive! The Catholic Church officially absolved Galileo of his 1634 misdeed in 1992 – the monk's status is still under review.

We are aware that much written here will be considered crazy by most steeped in our current conventional wisdom of science. But, the alarmed tenured should reflect on the fact that before Einstein formulated $E=mc^2$, everyone unquestionably knew that matter and energy were two different things, and that gravity was a force, not the warping of space.

Summary

Twain's insight is a wonderful prelude for understanding Einstein's mindset. Their stories will cover a broad horizon of knowledge. The wily paradoxes of medical statistics will show the daily problems that face physicians. History buffs will learn how British secret intelligence led to Einstein's greatest discovery. We'll explore Relativity's central themes, and appreciate how Einstein revitalized the Pythagorean philosophy that the beauty of math can bring one closer to the divine. Philosophy professors wondering why some types of mathematics were once taught in their classes might conclude that another renaissance is worthwhile.

Finally, politicians and their handlers will sleep more soundly. No need worrying whether they can justify why they chose a particular side of a political issue. Twain's World has ample statistics to espouse the righteousness of either side.

"A man can't prove anything without statistics; no man can...why statistics are more precious and useful than any other thing in the world, except whiskey - I mean hymnbooks."

- Mark Twain

Chapter 1

Twain's World: A Researcher's Nightmare, a Politician's Utopia

This chapter will meander like an old riverboat navigating the sandbars as it steams up the calmer waters of the Mississippi River. We'll take a variety of short excursions into Twain's World along the way. Our theme will be Twain's insightful caution about Lies, Dammed Lies and Statistics – not to mention common sense and pernicious conventional wisdom – as we quickly narrow our focus to medical issues. The uncertainty in medical statistics will be a sharp contrast to the exactness found in the physics of later chapters. Twain's view of human irony will show why you must squint and check your wallet when scientists try to sell you the latest modern version of old riverboat snake-oil, or politicians try to impose new medical regulations.

Politicians love statistics. A good example of a politician's favorite analysis tool is UC Berkeley's 1970s attempt to check for sex discrimination in its graduate schools admissions. Berkeley had no shortage of in-house talent to conduct this analysis and decided to use what it considered to be the reasonable criteria of comparing the ratio of females accepted over total female applicants versus the ratio of males accepted over total number of male applicants. Data showed that the ratio of females accepted was smaller than the ratio of males. Hence it was concluded that women were clearly being discriminated against. The next step was to root out this evil injustice by finding out which departments were most flagrantly conducting this sinister deed. However, the results of the individual departments made no sense – *every* department's male ratio was smaller than the female ratio. Wow! Politicians rejoice!

With actual data and honest analysis, opposite sides can come to a "factual conclusion" that serves their purpose. George W. Bush can travel the campaign trail and lament how every department in Berkeley has been proven to give excessive favoritism to women, while Hilary Rodham Clinton can solicit generous contributions from women's groups by vowing to eradicate the proven bias of universities, like U.C. Berkeley, that have unfairly discriminated against women. Both can passionately believe they are right. So, who is lying? Well neither of course – that's why statistical facts are so much better than regular lies! Statistics are solid facts compared to other campaign claims. Why hell, the term **solid facts** does not give statistics their saintly due for determining justice. They're miracles for lawyers making a healthy living arguing statistical facts to juries!

We'll soon see how this glorious reality for politicians and lawyers has a real downside for the FDA and medical professionals trying to decide if a new drug for cancer therapy should be used. Why do contradictory statistical results occur using the same data and analysis? How often does this happen? Before delving into these two questions – another more disconcerting question arises. Can statistics get worse than the dilemma above? Answer: Yes. Much worse.

While we look at the above paradox, we're going to step away from the specifics of the Berkeley study.[1] Evaluating a statistical study of sex discrimination is overly passionate for our purpose. The so-called "Battle of the Sexes" has waged on in many forms since before the trappings of civilization. A few examples of never-ending "issues" can start with the exclusively female population of ancient Lesbos while the Spartan warriors stole young boys from the poison of loving mothers; to the more modern *Taming of the Shrew* and the 1970's bra burnings. We'll surely be distracted by headaches from this eternal battle if we try to include political fodder within an objective study focused on this particular paradox. Instead, let's have a make-believe parallel study about eliminating headaches with aspirin or Tylenol.

---

[1] More detail on the Berkeley paradox is mentioned in the book, Gotcha: Paradoxes to Puzzle and Delight, Martin Gardner. Appendix A will touch briefly on federal guidelines for discrimination statistics and how easily they become corrupted.

A Simple Example of This Paradox

Let's ascribe the Berkeley male data to Tylenol and its female data to aspirin. Those accepted into graduate school will become headaches that disappeared in the study. Graduate departments become clinics and now our substitution is complete. In other words, _every_ clinic will show that aspirin is better than Tylenol, but when all the clinics are combined, Tylenol works better than aspirin! So which really works best?

In this numeric example, we will have only three small clinics conducting this aspirin/Tylenol trial. The volunteers will choose their pain reliever and note whether or not their headache goes away after using either aspirin or Tylenol. One clinic is in Aspen, Colorado, another in Vail, Colorado, and the third is in Mammoth, California.

The results in Aspen were 10 out of 22 taking aspirin said they got relief while 6 out of 14 volunteers said they got relief from Tylenol. We'll put this and the other clinics in standard chart form:

	ASPIRIN	TYLENOL
ASPEN	10/22 (.455)	6/14 (.429)
VAIL	6/9 (.667)	9/14 (.643)
MAMMOTH	13/21 (.619)	31/52 (.596)

Aspirin works best at _every_ clinic, since 10/22 is greater than 6/14, 6/9 is greater than 9/14, and 13/21 is greater than 31/52. So, aspirin will obviously beat out Tylenol when *combining* these boring trials. Right?

	ASPIRIN	TYLENOL
TOTAL	**29/52 (.558)**	**46/80 (.575)**

Wrong! All together, 46 out of the 80 volunteers choosing Tylenol said they got relief while 29 of 52 volunteers choosing aspirin said their headache disappeared. Surprisingly, .558 is smaller than .575, so Tylenol works best when combining these studies! Hmm...this doesn't make much sense.

So, these statistics are contradictory. Unless you have a financial interest in Tylenol or Bayer, or a bad headache, who really cares? Let's raise the stakes and make it personal. What if you have just been told you have a very aggressive type of lung cancer? Your doctor presents you with two treatment options. Neither option offers you much hope of surviving the year with your lung intact...AND you desperately avoid even acknowledging – you are likely to die soon from this cancer. What if the clinical survival results for your two cancer therapy options are exactly the same as those in the above aspirin/Tylenol trial? Do you choose the cancer treatment that worked best at every clinic or the one that worked best overall? Our college education tends to focus on only the benefits of statistics, which churn out graduates who believe such paradoxes are impossible. Statisticians quietly hide unsightly realities.

This illusion is also helped by our lack of teaching an adequate understanding of fractions. This statistical paradox will never occur with standard algebra addition where fractions are added by first finding a common denominator then adding the numerators over their common denominator. These statistics skip the common denominator part, and add numerator to numerator and denominator to denominator. The integrity of each clinical group size is not lost. This sort of addition is very common in the world of physics in which vectors are often added in this manner to get accurate results.

Our discussion on the different methods of adding fractions isn't doing much to save you from your cancer. So let's get back to the problem at hand – Which treatment is your better choice?

Answer #1

We'll combine all the studies. This is based on the reasoning that the more data you have, the more confident you can be with your results. Looking at all parts as a whole is best.

Answer #2

It's best to choose the one that worked best in each clinic. Since Mammoth had the largest patient base, it overshadows the Aspen and Vail results. Vail may have done something different that made both options more effective, while Aspen did something to decreased the effectiveness of both options. Neither should be penalized for having smaller groups. Mammoth unfairly overshadows the other trials. Disregard accumulated results contrary to this, as they are tainted. Use this logic to choose your cancer option.

Both answers sound reasonable, but which carries more weight? And are there more problems with the supporting arguments than we're addressing? Sure. For example, if you prefer the argument in Answer #2, what do you do when one worked best in 4 out of 5 clinics, and still performed worse overall? Do you still favor the reasoning of Answer #2? Clearly, these sorts of statistical problems make for difficult choices and sleepless nights for cancer patients and doctors trying to give the best care possible.[2]

And we can't even come close to touching on a fraction of other possibilities – what if there were further layers of contradictions? What if the best option for a postoperative largely intact lung for better breathing capacity might mildly contradict what's best for patient survival in a similar peculiar unsure way? What then?

Well, at least we know the answer to one brainteaser. Which job is easier: Problem Identifier or Problem Solver?

One can understand why mathematicians tend to migrate to the clearer world of physics, where laws like "The Conservation of Energy and Momentum" make math results from experiments exact, predictable and easily repeatable. Finding answers to medical predicaments with statistics often gives researcher, regulator and medical professional only regrettable choices, particularly when a trade-off must occur. Like Dorothy and the Scarecrow choosing the right path to Emerald City – statistical results often leave lingering doubts if our results are really leading us in the right direction. Unsure statistical analysis is often _the_ factor for making life-or-death decisions.

And the uncertainty gets worse. There are many other medical paradoxes and human intrigues that exist even before statistics are used for evaluation. Human factors can exponentially influence the uncertainty in medicine. Statistical analysis is supposed to be the unbiased arbitrator determining which medication is best for us. Unfortunately, this arbitrator is fundamentally flawed as illustrated by our first very simple example.

Some Medical Stories from Twain's World

A little anecdotal history of why U.S. medicine has changed its regulatory rules and statistical guidelines helps in understanding a major problem in medicine – Seemingly great new therapies aren't always better.

---

[2] Math hounds can proceed to the second half of appendix A for the beginning of a more technical math discussion in quantifying this paradox.

A good example is the use of oxygen tents for healthy newborns in the 1920-30s. There have always been a certain number of stillbirths or listless newborns. The thinking was that the detachment of the umbilical cord followed by a delayed birth, resulted in suffocation, and this was a major cause of these tragedies. Perhaps giving babies a nice dose of oxygen ASAP would help. Obviously, somewhere between death and a healthy baby lie those that barely survive and are left with permanent damage from lack of oxygen. Oxygen was first administered to the bluest of blue babies by putting the baby in an oxygen tent -- with great success. These heartening results led to the thinking that even normal births would benefit from an enriched oxygen environment after the trauma of delivery. After all, the elderly with lung diseases get great relief from enriched oxygen. What could be the harm? The logical benefit should be a healthier, smarter child. This therapy was new and somewhat expensive, so it was used primarily by those of wealth and medical knowledge. Unfortunately, healthy newborn babies are highly sensitive to excessive oxygen in the atmosphere. An over enriched oxygen tent can lead to blindness or mental retardation for a healthy newborn.

This VIP baby oxygen treatment had similar results to the VIP surgery given to President Lincoln for the bullet lodged in his brain—both led to tragedy. Some modern brain surgeons now believe Lincoln would have likely survived Booth's assassination attempt if overly-ambitious doctors had opted not to remove the bullet. President Andrew Jackson lived a long and healthy life having a lucky bullet lodged in or near his heart – a memento from a youthful duel. The young Jackson was lucky he had neither fame nor fortune to entice surgeons to perform VIP surgery. Likewise, parents wanting only the best for their child did not get what they had bargained for when paying for an oxygen tent for their healthy newborn. History has many analogous examples of logical, promising new therapies that were tragically wrong. Math – being the distillation of logic – can be deceptively poor at predicting these tragedies. Embracing a new therapy in medicine has its risks.

Alternatively, a delay in new therapy can be equally tragic. Patients are denied a new prescription therapy prior to FDA approval. Therefore any drug approved *and* found useful, can **always** be critiqued as coming to market too late...in *hindsight*. To skirt unfair finger pointing at the FDA, our example will be aspirin's ascent from lowly headache-reliever to a major component of heart therapy. Aspirin was already sold over the counter, so the FDA played the role of bystander rather than its usual role as gatekeeper as aspirin slowly became a mainstay for preventing heart attacks. Our story starts in the early 1960s, when a premier lecturer in Wisconsin's Medical School predicted that aspirin would soon become a common tool for heart therapy.

But this view was soon overshadowed by news of a rare but dangerous reaction to aspirin leading to Reyes Syndrome in children. The FDA was even considering removing aspirin as an over the counter medication because of this danger. By the early 1970s, numerous impressive papers presenting the virtues of aspirin for heart care added to the confusion. A general conclusion was gradually made -- aspirin is a good option for adults with heart concerns but risky for children. Unfortunately, aspirin therapy was advised by only a small minority of moderately risk-taking cardiologists in the early 1980s, and did not achieve broad consensus for heart care until the late 1990s. How many adults with heart disease would have led healthier, longer lives if aspirin therapy started earlier?

Actually, that rhetorical question has an answer that can be estimated. Assuming a conservative 50,000 deaths from heart disease annually and a third more from strokes will total to over 65,000 annual deaths. Now add in double this number for those who did not actually die from a heart attack or stroke but had a dramatic loss of quality of life from a weaker heart and you get close to 200,000 Americans annually. How many years was the delay of aspirin therapy and what percentage would have actually benefited from aspirin therapy? Let's give the delay of therapy a range of 15 to 25 years and the percent who suffered loss of quality life from the delay of aspirin therapy a range of 20 percent to 60 percent. This gives a range at low end of 500,000 to a high 3 million Americans who most likely lost out on the timely benefit of aspirin therapy. Even the smaller number in our rough estimates is much greater than all the Americans who died in combat from WWI to the Iraqi conflict. De facto caution has risks too. A reasonable claim from our sloppy math could say that over a million Americans were likely denied a longer healthier life from the cautious delay of aspirin therapy.

In truth, this estimate is not as sloppy as one might think when compared to the more complex formulas with fancy symbols and names in sophisticated statistical analysis. We will compare this simple estimate to the more complex methods later in this chapter. "Pay no attention to the man behind the curtain," in the *Wizard of Oz,* has the usual flavor of clamoring to queries about statistics' inner-workings.

Another more poignant example of rethinking the risk of new therapy exposure is the medical battle against AIDS. Many cancers and other rare catastrophic diseases are attacked in a one step at a time manner for a number of pragmatic reasons. One reason is the accumulation of dependable data. Historically, using multiple compounds in a shotgun attempt leaves the researcher with little if any convincing results. Unimpressive results from reckless data will get no further funding for research in the medical community. This ultimately hurts the effort to find a cure. This one-step-at-a-time approach is very logical in a purely theoretical world. However, this pragmatic method is painstakingly slow for those desperate and in immediate need for a cure. The movie *Lorenzo's Oil* depicts this frustration and despair for those suffering from an incurable, catastrophic rare disease.

AIDS was initially a very rare disease and largely concentrated in a small segment of the U.S. population: gay men. The disease spread mainly through sexual contact, and initially, a rapid death was believed to be certain. While the disease was rare, a focused group of millions of men were clearly alarmed and motivated by this immediate threat to their lives.

Imagine the feedback awaiting a medical researcher attending a meeting of gay men in the early 1980s to explain how medical research pragmatically attacks such rare diseases. He would probably be amazed or easily offended by comments like, "Are you crazy!" or, "Is the medical research community full of morons, or do you all just hate gays?" He might exit the meeting frustrated at their lack of understanding pragmatic medical research - unless he happened be gay. In that case, he would have likely started the meeting with, "Hey guys, we got a big problem with medical research moving too slowly." It's amazing how enlightened paradigm shifts to some are nothing more than the obvious to others.

The Problem: Why the FDA Can't Win

If the FDA is more conservative in risk-taking when exposing patients to new therapy:

- More research money spent on expensive trials (which may be of questionable value).
- Medications become more expensive for patients.
- Therapy is delayed, harming patients via denial of medication.

If the FDA accepts greater risk to insure that new therapy is not unnecessarily denied to patients in need:

- Increase in adverse risk exposure to patients.

And on what must the FDA rely to make this tricky decision for new therapies? Statistics!

- Our first simple example has shown how statistics are inherently flawed.

The FDA has since modified its method of bringing new therapy to market, partly due to input from advocacy groups for AIDS and breast cancer. But how different is AIDS from other very rare and terminal or debilitating diseases with no cure, aside from the lack of public advocacy for most rare diseases? Is medical research still moving too cautiously? The answer is relative: Are you asking siblings of a man who died from AIDS in the 80s, or the siblings of an unfortunate oxygen tent baby, or personal injury lawyers suing Merck, the maker of Vioxx?

Other Current Statistical Problems

The Vioxx Dilemma: A Welcomed Statistical Warning? Or Denial of Good Therapy from Bad Statistics?

Vioxx and Celebrex are both very effective medications for those who suffer from joint pain and arthritis. They work in much the same way as antihistamines do, but on a different problem. Just as with allergies, arthritis is the body's inflammation process out of control. But this time it's the joints that suffer, not the nose. Out-of-control histamine can cause the body to itch and sneeze from allergies. Vioxx and Celebrex block a more complicated compound than histamine, called COX2, which is responsible for much of arthritis joint pain. COX2 basically tells our body to attack itself at the joints and elsewhere when it is out of control. Vioxx's selling point to doctors was how it is very selective at controlling COX2 as compared to older, less potent nonsteroidal anti-inflammatory drugs, such as aspirin and naproxen (Aleve). Because Vioxx is very selective, it was thought to be both safer and more potent than older aspirin-like compounds.

Experts believe that many types of heart disease also involve over-active COX2 inflammation. Essentially, the COX2 tells the body to attack the arteries, which causes dangerous plaque buildup. It was believed both Vioxx and Celebrex would be shown to be useful in preventing heart disease and colon inflammation, which is often associated with colon cancer. This expert logic was very sound. So began the boring clinical trials that would surely demonstrate how Vioxx lowers heart disease. Unfortunately, the headlines tell us that the results were not boring and just the opposite from what was expected happened - Vioxx seemed to increase the chance of getting a heart attack! It was promptly pulled from the market and the pharmaceutical giant Merck faces possible annihilation by litigation.

While these results were a surprise for many researchers, history has shown such surprises are not uncommon. The conventional wisdom for pulling Vioxx off the market was that the dose for arthritis might be too high to maintain a healthy heart. These sorts of scenarios happen often. For example, iron blood level manipulated either too low or too high from the optimum range is very unhealthy for the heart. This serves as a prime analogy for how excessive manipulation either way of optimum COX2 levels might increase heart disease. Other experts feel there might be other molecular counter-mechanisms at play. We just don't know one way or the other for sure.

Hey, wait a minute. What about inactive people suddenly becoming much more active? Maybe too active to the point that an already weak heart can't handle all the fun! Is this a big factor for these statistical results? If one is not aching and constantly in pain, then why not hop over to fancy restaurants with rich food, or super-size the great fast-food deal rather than warm up low-fat soup at home? Or perhaps other patients took up more vigorous exercise, more lovemaking...maybe they entered that senior-center 1 K fun run to beat a grumpy neighbor to the finish line and show him a thing or two! All these activities might end in a heart attack if one cannot handle all the fun.

What about other lifestyle or medication interactions causing these negative Vioxx results? Here's a likely scenario: Maybe some have taken aspirin for joint pain for years and their heart has enjoyed inadvertent therapy. But the joint pain continues to increase even with aspirin therapy to the point where the doctor prescribes Vioxx. Joint pain goes away as Vioxx replaces aspirin. Thanks Vioxx. Arthritis pain's gone - Stomach irritation from aspirin's gone – Whoops – Heart therapy is gone too - Heart attack ensues. Should Vioxx be blamed and therefore removed from the market for making people more active and/or avoiding heart medication? How often did shades of this occur? Where did the statistical studies account for these possible scenarios?

How much of the negative statistics correlates to:

1) Vioxx, at any dose, is bad for the heart – versus

2) The mandated high statistical therapy threshold for prescription medication places Vioxx into a heart danger region for some people, but half the dose is safe for the heart but dangerously close to failing to meet the fairly demanding threshold for arthritis efficacy - a flaw in existing statistical analysis for slow degenerative diseases – versus

3) Vioxx is no danger to the heart but voluntary lifestyle changes from feeling better are responsible for essentially all the negative statistical data?

When all the information is taken as a whole, how can anyone argue the problem is a pandemic fiasco? Nevertheless, that claim makes great headlines and is likely to prevail in the jury room. To lawyers suing the maker of Vioxx who want to hurl this book against the wall, take heart:

The above logic doesn't hold a candle to a grieving widow mourning silently in front of a jury for several days.

A closer look at modern statistics will expose technical problems with statistical methods that aggravated the Vioxx dilemma.

Acceptance of Red Wine versus Vioxx Statistics

What is fascinating is how the acceptance of statistical results depends so heavily on whether or not they fit a particular agenda. The media, tort liability lawyers and a few regulators appeared quick to jump on board against Vioxx. Statistical evidence that proved red wine is good for one's heart was accepted at a snail's pace compared to the negative Vioxx statistics. A brief review of these two therapy options will highlight society's fickleness towards statistical results.

Compounds in Red Wine

In 1979, fairly compelling evidence emerged suggesting that wine might prevent heart disease. Dr. Anthony St. Leger wrote a noteworthy paper showing that greater wine consumption in different countries correlates well to less heart disease (see the graph below). Was this a direct cause and effect, or were other factors at play?

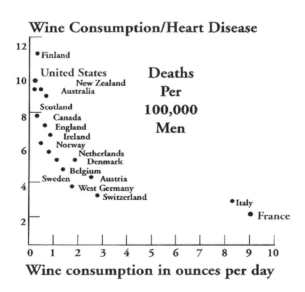

Many autopsy studies since 1904 had noted that alcoholics have very little plaque in their arteries. This is the same plaque that Vioxx was thought to prevent. This problem, called atherosclerosis, is a major cause of heart disease. A respected cardiologist published a paper in 1974 confirming the earlier autopsy reports indicating that moderate alcohol consumption probably reduces heart disease. By 1981, Dr. David Klurfield, *et al*, had completed rabbit atherosclerosis studies that showed modest heart protection from all forms of alcohol but dramatic protection from red wine in particular.

These sorts of results languished in obscurity for more than a decade until a Frenchman named Dr. Serge Renaud tackled the same issue with optimistic passion. He showed how the French have higher fat consumption, smoke more, and have higher cholesterol than Americans – yet have one-third the heart disease! And these statistical paradoxes continued as Renaud discovered that those in the French Provinces drinking more wine had lower rates of heart disease yet higher total cholesterol!

His work might have remained obscured had *60 Minutes* not presented the charismatic Renaud and his impressive results on national television in 1992. The public's passionate reaction was even a surprise to *60 Minutes*. Promoting alcoholism is a no-no.

This show was an absolute disaster for those who have had to deal with compulsive alcoholics. They knew this would be a perfect excuse for perpetual drunks to overindulge even more. Why, alcoholics could feel downright proud for making the healthy move from bourbon to red wine! No better example of "Lies, Dammed Lies and Statistics" existed in fervent anti-alcohol minds than this segment of *60 Minutes*. Many in the science community were accusing *60 Minutes* of again jumping too early to conclusions due to scientific ignorance. Not only that, ne'er-do-wells could garnish a wealth of logical conclusions from this troubling French data. Let's see, the French have much lower heart disease because of something in their lifestyle. What about their inclination for infidelity? A perfect excuse for spouses known to wander. Hmm ... "Dear, I didn't want open heart surgery while in office. I was just doing what was best for the nation." We don't think Hillary would have bought that line for a second. But we digress ....

So did **60 Minutes** make a mea culpa to calm the waters after presenting this troublesome French data? - and lower their ratings? Hell no! Instead, they ran a sequel presenting even more mountains of evidence supporting their theme that wine is very healthy for the heart.

So the genie was out of the vino bottle and wine being good for the heart was now in the spotlight. Yet most medical experts remained unconvinced by the data accumulated through 1992. But the wine studies just kept flowing. By 1995, a Copenhagen City heart population study had confirmed an earlier 1988 Harvard HRT (hormone replacement therapy) study having a study base of 80,000 nurses. And we can't forget the dramatic original St. Leger statistical population graph. All these studies concluded that wine helps prevent heart disease. Better yet, animal studies showed dramatic evidence that red wine prevents heart disease. Some researchers had replicated Klurfield's original atherosclerosis results with rabbits and other animals. Different animal studies were showing a strengthening and improved flexibility in the arteries from red wine. The red wine paradox appears to have paradoxes within the original French paradox. In both animals and humans, those imbibing greater quantities of red wine tended to have *higher cholesterol levels – but lower rates of heart disease!*

The following Venn diagram shows several population studies and is a good visual to see why population statistical results can be so slippery. Each circle represents all the variables in each study and red wine is just one of a myriad of possible causes for these promising results. A more realistic example to counter a red wine conclusion than the Clinton excuse can be easily found in the Harvard nurse study. There may be a bias – women who drink wine in the US are wealthier than the norm and therefore have better medical options, which may lead to a lower incidence of heart disease. If this thinking is true, then wine may have nothing to do with improving heart disease in the nurse study, but was just a marker for determining one's financial ability to get better health care. One can find, or at least imagine, factors in any one of these studies for reservations.

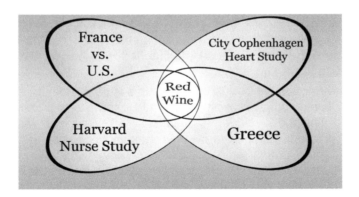

But the very existence of all these studies[3] cancels the legitimacy of many reservations. For example, infidelity vastly reducing heart disease only works for US versus France. Meanwhile high income skews in

---

[3] The alcoholic autopsy and several other population studies now completed beyond the Greece study belong in the Venn diagram, but have been left out to avoid visual clutter.

opposite directions for the Harvard nurse and French/US studies. Copenhagen City heart weakens the reasoning further that something other than red wine is entirely responsible.

Meanwhile the number of animal studies showing the benefits of red wine continued to grow. The only difference between two different groups in controlled animal studies (at least as presented by their authors) is that one group got the benefit of red wine while the other group did not. This is the major difference between controlled studies verses garnishing statistics from population studies. Controlled studies offer a form of triangulation confirmation for the uncertainty inherent in population studies.

An interesting controlled wine study originally started out as arguably the first study to definitively show how aspirin prevented heart attacks way back in 1970. Dogs' arteries were constricted to induce consistent mini heart attacks, and aspirin made these heart attacks go away – but stress counteracted aspirin's therapy in these studies. In 1995 Heather Demrow, *et al*, wrote a paper explaining how this same heart attack model tested wines and showed that red wines have several advantages over aspirin while white wine showed no benefit at all. One major advantage found in red wine was that – unlike aspirin – red wine worked even when the dog was stressed. Demrow's paper went on to explain why it is so important that heart therapy continues to work when humans are stressed.

Concurrently, a hospital in the USSR had completed a clinical trial of men who were recovering from a recent heart attack. It was led by Helena Lisunetz and showed that men who consumed the part of grapes that make wine red after their heart attacks had improved cardio function compared to those who did not.[4]

So one Russian clinical trial with humans is now added to a long list of other controlled animal studies, which show red wine prevents dangerous plaque buildup, strengthens arteries, and likely has advantages over aspirin in preventing sudden heart attacks...wow! All these controlled animal studies strongly support the many population studies. The grand total of all this pre-1996 evidence would surely convince even the most skeptical medical expert that red wine is good for the heart, right?

Wrong! In 2000, the above evidence was presented to a very active professor emeritus, once head of pharmacology, and still occupying his magnificent top-floor corner office overseeing a most prestigious university. His conclusion to all these statistical results? He remained

---

[4] Post heart-attack men consumed grape skins that remain in red wine vats for a while, but are immediately removed from the grape juice when making white wine.

completely unconvinced that any correlation between heart disease and red wine can be drawn! He felt any conclusion is premature until double blind clinical trials are conducted in the USA under more rigorous controls than in Russia.

As Twain said, "Denial ain't just a river in Egypt." Denial exists in the hallways of even the most prestigious of our institutions. Statistical hairsplitting can be a great comfort to those clinging to their old conventional wisdom.

The unconvinced medical expert countered that, while many studies showed averages favoring a benefit from red wine, some *failed* to reach the coveted benchmark of *statistical significance* – which in layman's terms means, "the stuff *really* works." He further claimed that wine consumption was likely to be proven bad for the heart. His reasoning was that red wine's tendency to raise cholesterol outweighs a decrease in heart disease! How reasonable was his thinking? Was it based on a strong anti-alcoholism sentiment? Did he feel an unconscious allegiance to his first gut reaction made in the early 1990s? Or is an insistence on a US clinical trial enough to comfortably consider all the combined evidence mentioned above as inconclusive? Clearly a favorable clinical trial conducted in the U.S. would add greatly to the evidence. But what are the strengths and weaknesses of clinical trials. Let's look *beyond* our aspirin/Tylenol problem.

Clinical Trials: A Good Statistical Problem Solver?

Human clinical trials are clearly the gold standard for medical research. They are the bottom line for new prescription therapy. But if they are so good, why do clinics with large group sizes have results so different when testing the same therapy? Why are so many new medications later taken off the market after spending millions of dollars on rigorous clinical trials that produce statistical results hailed by medical experts? The answers are many and varied. We will look at a few.

A great example of clinical trial "issues" is that 25% of men in a placebo (sugarpill) group of a Viagra trial said the pills worked great! This fact is either a great tribute to the power of positive thinking – a warning sign about clinical trial results – or both!

Another problem is if those conducting the clinical trial know which is the placebo verses the medication, they can clearly affect results through both conscious and unconscious manipulation. Therefore, trials are "double blinded" -- efforts are made to prevent personnel from knowing which pill is which. But how double blind are the studies once follow-up visits of pill x and pill y start showing a trend? Obviously the neat, cheerful, prompt volunteer is likely to respond differently than the disheveled, whiny volunteer that tends to reschedule visits. Once a bright nurse/doctor notices a trend in pill x versus y, he/she can greatly affect the trial's outcome by putting future volunteers into the right group. Also, how many paid "volunteers" actually take their medication? How many forget to take their pills until the day before their appointment with the clinic, so they double or triple the dose right before their appointment? How much does this vary from clinic to clinic? How many patients don't take their pills beyond a day or two but tell the nurse/doctor they have complied – just to avoid awkwardness -- particularly for the "paid volunteer"? How would these problems affect blood tests, therapy response and statistical results? Another curve ball: if a pharmaceutical company is contracting trials at several different clinics and finds that one clinic is achieving more favorable results than the other clinics – why not contract larger trials with the best clinic and avoid continuing trials with clinics obviously not monitoring the volunteers properly? This opens a whole new can of worms. Clearly our current methods of conducting clinical trials have some tremendous advantages over population studies – but they both have their own set of problems. Neither is a guarantee for surefire statistical results.

Finally, what are the math methods that determine whether or not statistical significance has occurred? What are some of the underlying assumptions? How reasonable are those assumptions?

Technical Problems Next Stop: Statistical Significance and the Binomial Assumption

(AKA: Good for Mathematics. Even Better for Twain's World!)

Statistical significance and most other statistical analyses rely heavily on the binomial assumption, which assumes a central clumping of events. The binomial assumption actually describes much of what occurs in medicine and the animal kingdom. There are truly interesting math theorems showing how clumpings that are not simple binomials can *sometimes* be manipulated to fit binomial parameters surprisingly well. A good example of what is and is not a binominal event can easily be learned from dice. The result of throwing just one die is not a binomial event. This is because any number between one and six is just as likely to occur – there is no clumping of events. If you threw a die one hundred times and divided the sum of all the throws by fifty, your result would be very close to 7. However, there is no clumping – no number is more likely to occur than another number. But throwing two dice is a binomial event. This is because the number seven is most likely to occur while the numbers two and twelve are least likely to occur. The following charts show how the throwing of dice can start out as flat, yet be combined into different binomials:

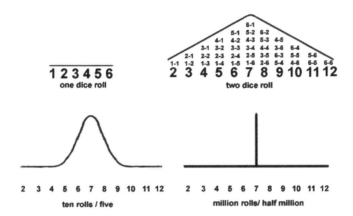

A number very close to seven is very likely if one were to throw ten dice and divide the total by five. This is a binomial event but has much tighter clumping to seven than does throwing just two dice. A million rolls of a die divided by half a million has a binomial spike at seven – rather than looking like a curve. Quantum mechanics statistics tend to have statistical spikes rather than curves because the measuring of events is often similar to a million rolls of a die divided by a half million. Medical statistics almost never have this level of statistical exactness.[5] The tighter the clumping of our data, the more we can confidently determine whether or not results are statistically significant when veering off normal clumping.

Mathematicians set parameters for when one should conclude that something other than random chance is most likely causing unusual results from binomial data decades ago. These statistical methods have been tweaked from time to time, but are still very similar to what was used more than fifty years ago. The threshold for designating something as out of the ordinary is usually pegged at less than a 5% chance (probability) of occurring under normal circumstances. This is most often denoted in life science papers as P<.05. This is the usual **statistically significant** threshold to prevent future false conclusions like those found by eugenics researchers over a hundred years ago. Researchers and the FDA utilize these reasonable parameters to determine whether a compound evaluated for prescription medication is indeed therapeutic.

---

[5] Math and paradox lovers see appendix B.

Obviously if a therapy is too hit-and-miss, why prescribe it? More important is the fact that most prescription medications end up having a downside for at least some patients and therefore, subjecting patients to new therapy that might not be helpful should be aggressively avoided. The method for evaluating new therapy as binomial data is straightforward.

A Typical Example of Medical Statistics

(*Note: Most non-math/statistician readers will find this overly technical and should just go to pg. 27.*)

This example will show how statistical results are typically calculated. This math is usually done by computer where the formulas and graphs we'll use are even further obscured from the user and the public.

Let's say a diet company has a new product that they think is better than its older formulation. The company had two groups with ten volunteers in each group take the new versus the old formula for three months. The new formula does have a slightly higher average of weight loss – but is it considered statistically significant? (Is $P<.05$?)

For each volunteer:

Weight loss on new formula, 6.2, 5.7, 6.5, 6, 6.3, 5.8, 5.7, 6, 6, 5.8 average 6 lbs loss

Weight loss on old formula, 5.6, 5.9, 5.6, 5.7, 5.8, 5.7, 6, 5.5, 5.7, 5.5 average 5.7 lbs loss

The equation often used is:

$$t= ((x\text{-}y) ((n_x n_y (n_x + n_y -2)/n_x + n_y))^{1/2}) / (d_x + d_y)^{1/2}$$

and $v= nx + ny\ \text{-}2$; where x is the average weight loss of the old formula, and y is the average weight loss of the new formula, and dx and dy are standard deviation (binomial clumping factor)[6] of x and y, and n are the numbers of volunteers in each group. The average of new formula is 6 pounds lost and the old formula is 5.7. The standard deviation from this data is .64 & .24 respectively. This gives us a $t = 3.03$. Then this t is then compared for the appropriate v on a table (see table below, v=18) that is

---

[6] This method uses an older standard deviation of the x group $\equiv ((x_1\text{-} x)^2 +...(x_n - x)^2)/n)$, where x is the mean.

based on results of random chance. The table shows P<.005 if t is greater than 2.878 on the table. Therefore this result is considered statistically significant with the P <.005 <.05 and one can reasonably conclude (beyond a factor of 10 of the P<.05 threshold) that the benefit of greater weight loss from the new diet formula is real and not a random chance occurrence.

One can continue to derive a degree of change certainty and other information – but we would like to turn our attention to the table for short cuts in showing trends. First, we will look at the equation in generalities.

t can also be written as:

$$t = \frac{\text{(difference in averages) (group size factor)}}{\text{(standard deviations-binomial factor)}}$$

The difference in averages is the primary factor. Standard deviation is the next largest factor, which has a binomial assumption that factors in the clumping strength of each group. We'll discuss this further later. Finally, the sample size is taken into consideration. Assuming group sizes, n, are the same, then (group size factor) goes to n as n becomes large. The chart also shows that for smaller group sizes the t is much larger for less than ten and the chart starts demanding "no-brainer" results for groups under seven in size.

The table shows the smaller data size again working against finding a statistical significance as the t starts at 6.314 and goes to 1.645 at infinity. Both n being greater than seven (v>12) makes the table's numbers look more linear than exponential. Are small samples really treated fairly? If the boundaries show such bias, what group size is really optimal? If we are *just* looking for reasonable certainty of change, are groups of 1200 really that much better (i.e. not concerned with beta factor) than, say, groups of 6, 12, 48 or 96 with respect to documenting a statistical significance?

The fact that infinity is not zero is full of philosophical irony.

For more math fun, let's look at a total data of three. Two from one group have data numbers a & b and the second group just has c. Therefore the equation:

$$t = ((x-y)((n_x n_y(n_x + n_y - 2))/(n_x + n_y)^{1/2} / (d_x + d_y)^{1/2}$$

becomes $t = (1/3)^{1/2}[(a+b - 2c)/(a-b)]$

ν \ P	.10	.05	.025	.01	.005
1	3.078	6.314	12.706	31.821	63.657
2	1.886	2.920	4.303	6.965	9.925
3	1.638	2.353	3.182	4.541	5.841
4	1.533	2.132	2.776	3.747	4.604
5	1.476	2.015	2.571	3.365	4.032
6	1.440	1.943	2.447	3.143	3.707
7	1.415	1.895	2.365	2.998	3.499
8	1.397	1.860	2.306	2.896	3.355
9	1.383	1.833	2.262	2.821	3.250
10	1.372	1.812	2.228	2.764	3.169
11	1.363	1.796	2.201	2.718	3.106
12	1.356	1.782	2.179	2.681	3.055
13	1.350	1.771	2.160	2.650	3.012
14	1.345	1.761	2.145	2.624	2.977
15	1.341	1.753	2.131	2.602	2.947
16	1.337	1.746	2.120	2.583	2.921
17	1.333	1.740	2.110	2.567	2.898
18	1.330	1.734	2.101	2.552	2.878
19	1.328	1.729	2.093	2.539	2.861
20	1.325	1.725	2.086	2.528	2.845
21	1.323	1.721	2.080	2.518	2.831
22	1.321	1.717	2.074	2.508	2.819
23	1.319	1.714	2.069	2.500	2.807
24	1.318	1.711	2.064	2.492	2.797
25	1.316	1.708	2.060	2.485	2.787
26	1.315	1.706	2.056	2.479	2.779
27	1.314	1.703	2.052	2.473	2.771
28	1.313	1.701	2.048	2.467	2.763
29	1.311	1.699	2.045	2.462	2.756
30	1.310	1.697	2.042	2.457	2.750
40	1.303	1.684	2.021	2.423	2.704
60	1.296	1.671	2.000	2.390	2.660
120	1.289	1.658	1.980	2.358	2.617
∞	1.282	1.645	1.960	2.326	2.576

Which is 6.314 < t = $(1/3)^{1/2}[(a+b - 2c)/(a-b)]$ where both numerator and denominator are the positive of their values (i.e. absolute) for P<.05. Now let's say b=c=0. So t= $(1/3)^{1/2}(a/a)$. No matter how large a is, it is not big enough to make the groups have a difference that is statistically significant.

Let's say c=0 but we play around with changing a and b, then 6.314 < $(1/3)^{1/2}(a+b)/(a-b)$. What if a and b are 10 and 8.3 respectively? Then the two different groups are statistically significant as is 1.0 and 0.83. But 10 and 5 as a and b are NOT statistically significant when c = 0! Hmm.

Boundaries in math parameters are always suspect, but it is self-evident from the table and its boundaries that we are dealing in estimates – not exactness. These results have been shown to be, at the very least, occasionally *very* counterintuitive.

We will soon see how unexpected problems occur even when using group sizes of ten, which is a common initial group size for clinical trials whose results are near the *middle* of the chart on the previous page.

Summary of Typical Statistical Analysis: Useful? Yes, but Not Perfect – Use with Caution.

Remember when we admitted to using sloppy math to estimate that about a million people likely suffered from the delay of aspirin heart therapy? Let's compare this typical statistical analysis to our sloppy and cavalier aspirin estimates. The infinity boundary on the table implies one can pick an infinite sample from both groups, have a difference in their averages, and STILL not be certain if you have a statistical difference between the groups – can't get much more counter-intuitive than that! At least the criticisms of our aspirin estimates were that they were rough estimates - not based on statistics that can be shown to have counter-intuitive areas...but...a question...will this book offer an obviously better method of statistical analysis for this diet formula example? Answer: No.

An Easier Example of Statistical Significance

A simpler look along the five percent probability boundary of statistical significance is the flipping of a coin. This gives the non-statistician a practical sense of when unusual events are finally determined to be statistically significant. If we flipped a coin five times and all five flips of the coin landed on heads, we would have had a series of events that together has slightly less than a 5% chance of occurring, and conclude that probably something unusual rather than luck is causing this unusual result. Four flips all ending up being heads is not enough flips to reach the under five percent threshold.[7]

The Reality of this Reasonable Boundary for Life Sciences

These boundaries of statistical significance are looked at differently by different industries. An intelligent life-science statistician believes a coin that is flipped four times and comes up heads all four times is not

---

[7] This is because under the norm there is a 50-50 chance of either heads or tails on each flip. The chance of two flips both resulting in heads is one in four (25% = P=.25). A general equation describing the chance of a coin always turning up heads is $(1/2)^n$; where n is the number of times the coin is flipped. Four flips all heads is a one in sixteen chance of occurring while five flips all heads is a one in thirty-two chance. (Chance of four flips all heads=1/16 chance = 0.063 > 0.05=5/100 = 5% = (P=.05) =0.05 > 0.0313 = 1/32 =chance of five flips all heads. Note: There is excessive repetitive nomenclature in statistics.)

statistically significant because of his industry standards. However, there were seedy casinos during the 1950s that would have greeted our statistical expert with open arms. The pretty casino girls would have escorted him right to one of the tables where they can offer playing customers free drinks. Ten bucks says our statistical expert would do very well at the free drink table – at least during his first visit, when the casino wants to get him hooked. But it would have been all downhill from there – never showing a statistically significant anomaly on how he became penniless. Unfortunately, this statistical problem doesn't stop with naïve suckers visiting fixed tables. Subtle manipulation can easily occur everywhere.

These reasonable parameters of statistical significance can impede good science in a variety of subtle ways – if their shortcomings are not acknowledged. The flip side to clever manipulation is how a statistical parameter like the reasonable insistence of a 95% certainty ($P<.05$) of therapy benefit has a not-so-obvious downside of missing useful therapy options. Researchers can easily miss a 30% therapeutic improvement. Hence many compounds that show modest therapeutic possibilities will be rejected as possible therapy options while only those with dramatic results are likely to make it through initial small group screenings. This will be shown by computer programming.

Some argue that the insistence on a 95% threshold helps perpetuate medications having a downside since initial screenings insist on a dramatic improvement. One can reasonably assume that the medication is affecting more than just the malady in question and this correlates to likely negative side effects. Should the 95% threshold be changed? This is a polarizing question. Some experts are inclined to increase therapy certainty to say, 97%, thereby increasing the likelihood that the benefit grossly counterweighs inevitable adverse reactions. Others see the problem as completely the opposite and counter-argue that a 97% threshold would only increase the good-bad spiraling scenario. Those who suggest a lower threshold, accompanied by higher toxicology thresholds, believe this would give doctors the option of milder long-term improvement with a smaller chance of adverse reactions. This different guideline concept for drug approval might offer better therapy options for some slow degenerative conditions, such as arthritis and multiple sclerosis.

This debate becomes even more polarized by charlatans and quacks making outrageous claims for their products and insisting that a FDA-BigPharma conspiracy denies the public good therapy options. Quacks love anecdotal evidence and claims such as "50% report marked improvement," "help the blind to see," "prevent AIDS," etc., yet they lack credible evidence to support those claims. A popular radio and TV doctor often laments how many lawyers and nutritional peddlers deceive through phony science. He is a true believer that one is taking unnecessary risks when not waiting until well-established conventional science has finished its pragmatic evaluation – good advice – as noted in our look at VIP medicine. He has personally seen and heard mountains of anecdotal stories of how the desperate sick can buy into unsound science - and even reject a good well-established therapy option - in favor of the witchdoctor who magically removes chicken parts from their innards. Meanwhile the curable cancer tumor continues to grow to become an incurable problem.

This radio personality enjoys informing his lay audience about recent scientific papers. One of these paper's statistical conclusion was how the female body changes in weight and form (bloating) during the menstrual cycle. Guess what? The paper concluded that there was no change of statistical significance. Hmm...just more proof that wives complain to their husbands about nothing? Or, is it in the realm of possibility that this study's conclusion is a good example of statistical shortcomings in medicine? The changes were too subtle, unpredictable and random to reach our statistically significant benchmark. Maybe women do know more about their own bodies than do fancy medical statistics!

The use of fish oil to combat heart disease is another good saga in which studies failed to show a statistical difference benefit because the results were often too modest and varied too much from person to person. It took way too long to reach the right conclusion that fish oil, by in large, helps people with heart trouble. Our current statistical parameters and evaluation methods were simply overly-restrictive for this therapy option.

Examples of the high statistical threshold problem with auto-immune diseases are found in the therapy options for multiple sclerosis (MS) as well as for arthritis. A recent example is the suspension of multiple sclerosis (MS) medication that appeared to be very effective at impeding the usually very slow progression of MS. However, its overly anti-inflammatory potency occasionally allowed for the dangerous invasion of viruses into the brain. Would lower efficacy thresholds prove more effective overall? Would a less potent, yet safer, anti-inflammatory coating of the cerebral vascular linings be a better option than this apparent over-potency? Are our efficacy statistical methods originally designed for acute maladies appropriate for these types of ailments? Was this part of the problem with Vioxx?

We'll do a little number crunching to get a handle on a known improvement that does not reach efficacy thresholds of $P<.05$. This will show a shortcoming of statistical significance that can cause misleading results for fish oil, monthly bloating studies, and likely miss good MS therapy options. This flaw is rather easy to expose from a mathematical viewpoint. It's a little disconcerting that the medical research community is, by in large, oblivious to this problem.

Cookbook Overview of the Detail Showing a Relevant Shortcoming of Statistical Significance

(The non math/technical should just go to the next paragraph.)

Most Microsoft Excel programs have both a random number generating program and basic statistical analysis programs that can be used to test and appreciate statistical limitations. Create a set of random numbers between one and ten and organize in groups of ten (i.e. ten patients per group). To half the groups, add 1.5. This gives a 30% increase (with regard to the mean being 5) for the changed groups, compared to other groups that are not increased. Now, separately for the changed and unchanged data, add two groups together and divide by two to get a binomial curve similar to the rolling of two dice. Now run statistical comparisons on the groups increased by this type of underlying increase of 30% versus those with no increase.

Results from a Known 30% Improvement

The results of the cookbook method above, which made groups with a 30% known improvement, will show that less than 10% of the comparisons will have a statistically significance ($P<.05$) difference with the one-roll-of-the-die data, and a less than 20% chance of statistical significance difference with the two-roll binomial groups! Reviewing human and animal raw data of many maladies shows that the binomial assumption is a stretch, and rather multiple mini-clumping is reality.[8] This makes our binomial assumption even less accurate. Math morphing and our tables can work even further against finding a statistical significance when, in fact, a meaningful difference exists. So the 20% chance of finding a statistical significance from a 30% underlying improvement could actually be high! And it gets worse with human mischief.

It can be argued that it is an advantage for some government funded institutions to find promising, but not yet conclusive results. That way, the funding continues on an already well-established issue in which the institution is the recognized expert.

---

[8] For example, atherosclerosis in both humans and non-genetically altered animals.

What do Moderate Cookbook-like Results Get Us?
Recognition, Envy and Intrigue...or a Steady Paycheck?

Let's say you're a researcher who had good results showing a statistical difference in your mouse or rabbit trial. Someone conducting the same trial to try to verify your work would most likely not be able to reproduce your impressive results if the real improvement is only 30%, as described above. In other words, they have less than a 1 in 5 chance of also obtaining a statistically significant result. Now who to believe?...Egos can be publicly displayed in transparently cloaked civility of peer review articles and letters to the editor - like gossip of those lacking proper attire whispered about at a cocktail party. Was the original paper authored by exaggeration or data cheating? Or is the rebuttal author to the original paper technically sloppy, or incapable of conducting a test to the rigorous standards of the original scientist? What about the wrath of envy from those in your field who were not the first to discover this therapy option? Do they have control over your future funding? Scientists have learned to ease this problem by making less bold conclusions. So, promising but inconclusive results can actually work to a researchers advantage. Delayed findings mean prolonged funding. Simply make a low-keyed conclusion that one cannot draw a definite conclusion from these promising yet inconclusive results. If the data is similar to our random generated improvement above, decades later they may still be coming to the same conclusion – more studies are needed. There exists a financial incentive to maintain the status quo statistical methods. A scientist has traded a brilliant discovery badge of honor for the more valuable steady paycheck in these scenarios.

The Selenium Example:

A study of selenium therapy is a good example of how the desire to continue research rather than introduce new therapy can exist.

Comprehensive cancer clinical trials were instigated when a correlation between high levels of selenium in local produce and a lowering of prostate and skin cancer was shown several decades ago. The National Institute of Health (NIH) funded a large set of clinical trials of selenium effectiveness. Unfortunately, it wasn't known what type of selenium and how much selenium would work best...if it even worked at all! So the researchers made reasonable attempts to estimate effective dosages and used a common form of selenium found in yeast that also protected bacteria cells from becoming cancerous. The selenium found in yeast that also showed the most promising results in cell mutation studies was selenomethionine. The daily dose was pegged at 200 ug. The results were remarkably good even with these bold "stab in the dark" parameters. The onset of both prostate and colorectal cancer was lowered by over 50%. Lung cancer also showed a dramatic improvement ($P<.01$, $P<.03$, $P<.05$ respectively). So selenium therapy is right around the corner for cancer prevention...right? Wrong.

After reviewing the clinical trial results, a lower dose of a different type of selenium would have likely produced even better results. For example, well over 50 liters of red wine must be consumed daily to get the daily dose of 200 ug of selenium from the bottle of red wine containing the highest level of selenium from *all* of the many red wines tested from around the world. Also, animal and cell studies have shown that the types of selenium found in red wine work much better against cancers than does the selenomethionine used in the clinical trials. There are several other technical reasons beyond the wine data that also conclude that the selenium used in this clinical trial was not close to being ideal. Hindsight is almost always so much clearer than foresight. There is an optimal level and type of selenium; we just don't know what it is.

Again, prostate and other cancers were lowered by over 50% using the wrong type of selenium complex at a wrong dose for optimum cancer prevention...so...why is therapy being delayed? We doubt this would be the current status of selenium therapy if Bill Gates was overseeing the selenium project for NIH. Mr. Gates built an empire by offering the best currently available product for his customers while continuing to build something better. Should thoughtful selenium therapy really be delayed for decades while research continues?

Some of the proposals from research institutions are to conduct more thorough clinical trials for selenium on a more massive scale. A patient study base as high as 34,000 patients has been proposed. The sales pitch for such a large, publicly funded trial is obviously getting more useful information from a larger patient base. However, huge random samples almost guarantee illusionary trends. For example, flipping a coin a thousand times almost guarantees that these flipped coins will contain one or more sets of five consecutive flips that are all heads – or $P<.05$ for these five flips – something of statistical significance that deserves further study. Likewise, one is almost guaranteed with 34,000 selenium patients to get interesting statistical trends both real and statistically imaginary...not bad....if your economic payoff comes not from getting a drug to market, but from public money research funding.

OK, it's easy to be a problem identifier for statistics – it's improving the usefulness of statistics that's tough. Any talented mathematicians who can mitigate these statistical problems will be welcomed with open arms – except by those who are trying to perfect perpetual government funding via imperfect statistics.

More "Issues"

We will continue to briefly describe a menagerie of disjointed medical intrigues. Every short story that follows is an ironic tale that some folks will want to either: over-promote, or hide. Statistics can build beautiful facades for either agenda.

Keep It Simple When Evaluating Medication

One of the wiser mathematical decisions was to evaluate only one active ingredient at a time in clinical settings. If one were to evaluate more than one, then the mathematical statistics become quite complicated with a much lower degree of confidence regarding its results. Ninety-five out of a hundred mathematicians would have approved of this practical parameter for evaluating and creating single active ingredient pills. However, even this decision has come back to haunt medicine. How?

Both the evaluation of one medication per pill and economic factors has contributed to the breeding of the super resistant bugs we are fighting today. From a mathematical practical standpoint, individual evaluation provides the clear data needed for good evaluation – so don't combine and muddy the waters. Not that long ago the medical community believed that one-compound pills had no real downside. A doctor can simply prescribe more than one medication if that's what's needed. No need for combo drugs - very simple, sound logic. Too bad that good logic of the 1980s aided the arrival of super resistant bacteria.

Big versus Boutique

Our economic great deal buffet at the Vegas casino equivalent method of medicine is also a primary culprit in aiding the arrival of resistant bacteria. How long would a 1980s HMO cost administrator remain employed if he/she allowed expensive dual medications when the insert for each medication clearly states that it alone will probably solve the problem? Subtle pressure will mount even as physicians try to explain their reasoning for this added expense. So in true gigantic HMO form, the cost administrator is guiding doctors toward providing more cost efficient care...a counter force to patient feedback...hmm...maybe it's best to skip the medical buffet line and go for the old fashioned sit-down dinner medical equivalent....if one can afford it.

The over-use of single-compound antibiotic pills, and the impersonal cost-effective streamlining of healthcare, combined to create the perfect growing conditions for resistant bacteria.[9] "Compounding" is one of the better examples of many solutions to preventing this problem. The traditional practice of "Mom & Pop" pharmacies combining different medications into one pill to help fit a particular customer's need is still around - but losing ground every decade. How much better are the very expensive new combo antibiotic drugs compared to a local pharmacists simply compounding for each patient?

If patients had a better personal relationship with a good pharmacist and/or general practitioner, there would have been a faster response when patients began to say, "Hey Doc, these pills ain't workin' like they used to." This would have at least impeded the arrival of super resistant bugs. There is no substitute for one-on-one communication between patient and a *good* caregiver – but even Huck would be quick to note, "It just ain't the cheapest method to dispense healthcare to kinfolk."

Also, which is better – a talented GP who knows you from several past visits, or, "Come on in, get a number, wait in line, and one of our many doctors will see you oh so briefly"? If a picture is worth a thousand words, what are actual previous visits with the patient worth as compared to reading the chart made by another? On the other hand, seeing several different talented doctors in a large HMO setting is like getting several second opinions, which clearly has a different set of advantages.

Government policy can pit the financial well-being of large care-givers against boutique medicine as they compete for profits or maybe just survival. Which type of medical option should be encouraged for Americans – large HMOs or boutique medical groups? It is very easy to craft statistical facts promoting either side of this debate. How much will government policy determine which is readily available? Looking at the medical policies of the U.S., Canada and Britain will show different pitfalls for each country's medical system.

Centralization of Medical Policy versus Diversity

Britain, Canada and France rapidly nationalized (socialized) their medical system along with steel and other ailing industries decades ago while the U.S. was making comparatively modest moves for government control in medicine.[10] Decades of national pride and an earnest desire for the best medicine possible have resulted in obvious good and bad anomalies in each system.

---

[9] There are numerous reasons for resistant bacteria, such as patients not completing proper dosage cycle, etc. The single active ingredient concept is just one of the many significant reasons for the unfortunate arrival of resistant bacteria.

[10] The interesting U.S. exception was the long-ago limitation of qualified students entering medical schools.

"We're in H.M.O.s—the closest to legit we've been in years."

Canada and Britain transformed the triage concept in emergency rooms to medicine in general. Get a ticket and stand in line – but it's free! The pay scale of physicians was set by government policy and not by the individual users. British and Canadian policy mandated that those over a certain age (65 or 70) be denied some types of therapy while these same therapies were widely available and common in the U.S.

The lines for so-called elective surgery were months and sometimes years long, but it was FREE![11] And it was distributed fairly of course; there was never any cutting to the front of the "queue" for those of power and influence...right? As the grumbling in the lines grew and the medicine dispensed became more mediocre, Britain took a turn toward privatizing part of its system while Canada maintained a steadier course of socialized medicine. Prime Minister Maggie Thatcher made changes in the UK's medical system so now the typical UK doctor can do the 9 to 5 on government time and then switch to patients willing to pay for after-hour services. Hence, doctors are semi-motivated to please and increase demand for his/her after-hour availability. Interestingly, Canada continued to quash this after-hour concept by making it illegal for doctors to give preference to those willing to pay-out-of pocket for after-hour treatment. Canada took the self-righteous moral high ground of maintaining equality for mediocre medical care. It's good to know that eliminating liberties to maintain equality is still the most noble of necessities in some places even after the collapse of the Soviet Union. Luckily, ninety percent of Canadians live less than 50 miles from the USA border. So their very selfish desire to get the best healthcare possible can be satiated by an hour's trek. Kinda 1920s Prohibition in reverse for the ailing sober. This Canadian law shows how the naked power of envy can grip any nation – even one that prides itself on personal liberties.

The flipside of modern socialized medicine is the current U.S. private/government hybrid system that synergistically enhances both the good and dysfunctional aspects of health care. Our chaotic non-definable system demands the liberties of choosing our options and sometimes paying out-of-pocket for the "best," but – hey – if we can pass the bill off to Uncle Sam or someone else...all the better! Bait and switch has reached a new zenith in American medicine: Generics being as good as non-generics; HMO's and insurance companies having contracts where you only **thought** you were covered; and doctors' bills are the obligation of someone else. Also, we can't forget multimillion dollar TV ads from a large pharmaceutical company asking you to ask your doctor if the purple pill might be right for you – without even saying what the purple pill does! And today's current grand prize winner for irony – Pfizer arguing that young

---

[11] Shh, pay no attention to how much you're paying in taxes ... and if you do, take heart. The rich are paying even more.

struggling middle class families should subsidize Viagra medication for wealthy old men in the name of not neglecting the elders' life-saving medical needs! Twain's dead-on wisdom, "Figures don't lie, liars figure," kicks into high gear with the aid of flexible statistics justifying opposing positions on how to change our medical system to *their* advantage.

Mark Twain Tobacco

One of the best paradoxical examples of "Figures don't lie, liars figure," can be learned from a famous weed once sold as a popular cigar called "Mark Twain Tobacco." How differently socialized medicine versus the U.S.'s more laissez-faire system dealt with tobacco is considered telling by some medical experts. The money being made and saved created a paradox.

Believe it or not, tobacco was initially touted as a health aid when introduced in the 1500's to Europeans – and for good reason. Tobacco smoke contains very high concentrations of nicotine, which besides being an addictive mood-altering agent, also happens to be a very good insecticide. Hence, fewer disease-carrying lice, ticks and mosquitoes meant a smaller chance of contracting the plague, yellow fever, malaria and a host of other diseases. Dying of lung cancer at some age beyond 60 was of little concern when most were lucky if they celebrated their 40th birthday. But tobacco changed from a health aid to a menace as life expectancy crept up and insect-borne diseases were brought under control by other means. Lung cancer and other diseases from smoking became major causes of early death and the US government was making great efforts to curtail smoking by the late 1960s. By the 1990s, tobacco companies were negotiating to reimburse states for the financial damages tobacco had inflicted on state-financed health care systems.

European state-run health care seemed to turn a blind eye to the dangers of tobacco. Why? Some have argued that the answer is that smokers were much less of a burden on state-run healthcare than are nonsmokers.

Question: How can this be?

Answer: Most people in Western nations either die of heart disease or cancer. Medical costs in the 1970s through the 1990s were quite high for heart disease. Lung cancer was a rather inexpensive way to die. Lung cancer chemotherapy agents were inexpensive. Usually, one set of chemo doses resulted in a regression that was usually very temporary and then the cancer would quickly return to aggressively terminate the smoker's life. Open heart surgery was much more expensive than lung surgery and the subsequent heart medication tended to work pretty well. Heart patients of the 1970s were much more likely to survive for decades and develop other maladies overburdening state-run medical systems even further than were lung cancer patients.

Some have argued that this is why many European countries were not vigorously opposed to smoking. If correct, then America's settlements by the tobacco companies to pay for excessive medical costs are statistical shams – these states actually *saved* money when people died of lung cancer from smoking! Let's see tobacco companies argue this thinking to a jury and see where sound accounting logic will take them financially!

America versus Western Europe: Same Statistics, Opposite Reactions

Let's be honest: Social Security would be flush with money if everyone would just die when they turned 65. *The* major financial factor for maintaining a fiscally sound retirement system is *very* contrary to what's best for the retirees. Smoking kept health and pension costs down for socialized countries in Europe. The fact is: nations with socialized medicine have often been too slow to embrace health policies that extend the lives of geriatrics. And the story gets worse for advancing medical research. Socialized countries feel obligated to play a face-saving catch-up to countries having the unfair advantage of not criminalizing private profiteers.

Profit and the Rich: Use the Filthy Rich to Everyone's Advantage?

Another up side to keeping a large part of our medicine privatized is the fact that the very rich are willing to spend tons of money to keep themselves alive. Nothing will circumvent stagnant "good old boy" peer review research barriers faster than the desperately sick rich. We have seen how VIP medicine is often a complete failure – so their money might actually get them to the grave faster than otherwise. It was no accident that very expensive open-heart surgery was originally headquartered in Houston, Texas, where rich, middle-aged men were praying that very risky surgery would let them drill more oil gushers. Fairly inexpensive procedures are now available to many because a few were willing to spend a lot of money to selfishly pioneer heart surgery. Likewise, if a five million dollar immune agent batching process is denied government funding for being theoretically unsound, and rejected by the business community for being too financially risky – "Who do I write the check out to and when can I get the first batch," may be the response of the very wealthy movie star with AIDS.

The Canadian healthcare system sees this opportunity of the rich as more of an evil injustice than an avenue that has advanced medicine...but...Bottom Line: The self-serving motivation of the wealthy to extend their own lives sometimes ends up helping us all – and hey – they take all the risks!

Profit Motivation: A Cure-all End-all?

The history of profit has proven time and again that rewarding results creates great benefit for all – and motivation to commit larceny for many. Twain loved tales of how the human passion for wealth is not always the quickest path to a cure-all end-all. We'll point out a few profit induced medical faux pas.

The Aspirin/Lecithin Pill: A Business Lesson

So many illustrative anecdotes hail from the humble aspirin. Take this tale of a good idea derailed by corporate profits and government regulation. Once upon a time, aspirin was only a pain reliever for the occasional headache. There was often minor stomach irritation, but that was simply the price paid for relief elsewhere. As the 1990s started rolling by, aspirin came into its own as a daily preventive heart therapy - bringing with it a potential for chronic stomach irritation. So a good-hearted savvy researcher at the University of Texas had a simple, effective, cheap solution – make a combo pill of aspirin buffered with lecithin – AKA Cremora, the non dairy cream made from soybeans. Lecithin can nip the irritation in the proverbial bud by coating the tummy as the aspirin/lecithin pill dissolves. This way the acidic aspirin pill is less inclined to burn a hole in the part of the tummy where it is dissolving.

This great idea was way ahead of its time and was guaranteed to be a success. Right? Wrong. The researcher was to see another side of the looking glass when he took his patent in hand to the pharmaceutical companies. The point of view of actual business: Where's the profit and what are the regulatory hurdles? Was his discovery a solution to a problem, or just a set of new headaches? Prescription sales of ulcer medication had become quite brisk since aspirin heart therapy became common. Does the large pharmaceutical company that controls the aspirin market also have Rxs for ulcer and stomach indigestion? Also, would Cremora reduce aspirin's therapy on headaches? Would different aspirins for different problems overly confuse the public? Is aspirin stable when mixed with Cremora? And the issues went on and on...hmm. Maybe the best option for a big pharmaceutical company would be to assign this project to someone in the company who is known to never get anything accomplished? By 2000, enteric-coated aspirin entered the scene to address the problem of stomach irritation. So, the Cremora option now faced brisk competition with only a year or two of exclusivity left from its patent. Enteric touts that the aspirin will dissolve in the intestinal tract where there are a lot fewer nerve endings, so irritation is less noticeable – yet the intestinal lining is less durable than the stomach. Hmm...this isn't simply sweeping a known problem under the proverbial rug, is it? To make a sad ending short: Our researcher has had to learn to live without a spare mansion in the Swiss Alps.

Moral of our story: Profit motivation is no guarantee that a good medical improvement will move forward.

Our last jammed-packed tale will be a make believe scenario of a small drug company that was glad the Cremora/aspirin pill never came to be. Let's say you're the proud owner of a small drug company whose flagship product is an anti-ulcer medication. You've spent your millions getting your drug approved for market and have been selling your product for years and are now making a comfortable profit on your investment. Your product is selling briskly particularly since more and more people are taking medications, like daily aspirin, that tend to irritate the stomach. There have been no complaints about adverse reactions to your medication. This is much better than your major competitor's anti-ulcer medication that is being removed from the market because of an occasional rare complaint of a major problem. But because of your competitor's rare reaction problem, the FDA is re-evaluating its approval methods and asking for industry input. Your input will of course be – "have very expensive and extensive clinical trials for new therapies to ensure the health and well being of the public" – which also happens to coincide with your company's financial health and well-being. More strict and expensive regulatory hurdles for new competition correlates to a longer monopoly for you and higher unit cost for them. Your company's luck is holding out as the FDA has adopted many of your suggestions – increasing the costs for a new competitor's medications.

But now the FDA is rethinking its policy at bullet-point-Internet speed:

- The FDA is receiving your competitors' input.
- Your company neglected to retain a good lobbyist with the pork-barrel mantra, "Help so few with just pennies from each of so many."
- And the FDA decides it's unfair and unwise in terms of public safety for your company not to conduct thorough trials costing additional millions.
- Your company is now haunted by what it perceives as excessive regulatory pressure/expense your company originally helped to instigate.
- As Congress wrestles with the concept of allowing a streamlined entry of prescription medications from foreign nations not burdened with such outrageous regulatory/safety costs.

- Congress wishing only to save the U.S. consumer money, pays particular attention to inexpensive imports from countries paying their workers 1/100 of USA wages.
- These countries are demonstrating even more capitalistic wisdom by quietly financing campaigns of likely U.S. presidential/congressional winners – surely such financial and political sophistication correlates to inexpensive, astute drug manufacturing.
- Why deny American consumers this great opportunity?
- Is the United States still a great country or what! It's nice to know some institutions remain rock solid from change since Twain's time – we still got the best congress money can buy.
- Too bad we lost good prose to bullet points.
- And how smart we've become to accept foreign currency at such favorable rates to our dollar – knowing our dollars' true worth when the U.S. is importing everything.

Bottom line of the above bullet points: A not-so-hidden game for the medical industry has become, "Convince the politicians and the FDA to make unprofitable demands on your competitor and not on you!" Statistics become pawns in the wars of costs, profits and market share for everyone from small companies to foreign nations.

Twain loved the opportunity and action of capitalism. But the world of free enterprise is rough and tumble – particularly when there are different rules for different people and hidden profit agendas masquerading as policy improvements for the good of mankind. The remarkable success of the United States' short history is in large part due to those like Twain, who made constant efforts to remove annoying flies found in our useful fuel of free enterprise.

So Far, So Good

Our country has relegated widespread famine and wholesale unemployment to our history books and foreign nations. How much wealth sharing versus motivating progress with wealth rewards is our country's primary, never-ending political debate.

But how motivated is our society in finding a cure for that disease lurking out there to get *you*? How do we maximize medical progress and maintain a reasonable degree of medical fairness? History shows that where envy dominates medical policy – mediocre medicine is soon to follow. History has also shown how overly centralized medical mandates can restrict discovery – alternatively – the sheer weight of government funding often speeds the discoveries for some cures. What's the proper balance?

Enough of the Profit/Political side of medicine!

Our whole discussion has a tone of being quite jaded against statistics and deception. But we are delivering counterarguments to today's naive awe given to statistical results. Our contention is that statistics still remain our best guide for *many* unknowns – but their shortcomings and ability to deceive need to be acknowledged.[12] The integrity of the presenter is what counts most in statistics...and just to keep us uncertain...even our jaded sarcasm about human trickery deserves a modest caveat. A little big-hearted deception can sometimes be useful as shown by a modern day Tom Sawyer.

Twain's Tom Sawyer had a knack for making a profit while keeping everyone slightly ignorantly happy. Our modern day Dr. Tom Sawyer at a large HMO is tuned into the fact that soothing patients' minds is almost as important as fixing their bodies. Dr. Sawyer's go-to solution for a fretful mom whose toddler has typical diarrhea was *not* to suggest that the problem usually takes care of itself – ergo, mom feels slighted by a condescending, uncaring doctor – *nor* does he give into mom's nervousness by prescribing a potent Rx antibiotic that tends to further irritate an already raw tummy. Instead, he would write on a prescription style pad a note to the downstairs pharmacy asking that they sell bismuth salicylate to mom for her toddler. This mild anti-microbial happens to also be very soothing for a tummy. He makes sure to tell mom that the HMO pharmacy downstairs has this in stock and it will only cost her the $5 co-pay with no annoying paperwork. He reminds her to call in two days if this medication is not strong enough as they will then more aggressively deal with the problem. Our savvy Dr. Sawyer has just made the HMO, Mom and her toddler happy. The HMO pulled a small profit on their $5 sale of generic Pepto-Bismol compared to losing money if Dr. Sawyer had given into the pressure to prescribe an unneeded Rx – Mom feels good in assuming this caring, unpretentious doctor has given her some "good stuff" not otherwise available to her – and her child has a more comfortable tummy as this soothing low-grade anti-microbial and nature solve the problem. Dr. Sawyer believes that the subtle art of deception does not always lack virtue.

---

[12] Some good books for further reading: *Introduction to Statistics*, Paul Hoel (good general technical reference); *How to Lie with Statistics*, Huff (old classic, easy-reading); *Rethinking the Foundations of Statistics*, Kadane, et al; *Advanced Statistics Demystified*, Stephans.

Improvements and Solutions: The Hard Part

Disease – Genetics – Medication – Diet

Here is a simple question: Why would medication that works so well for many, be terrible for a few? The obvious answer is people are just too different...duh. Medication is given to people not diseases. This brilliant revelation can be dissected further in the manner used to predict human behavior.

Psychology's concept of segmenting genetics and environmental factors shaping a personality is also useful for dissecting options to improve medication utilization. The presence of pre-existing medical conditions is one environmental factor handled very well in America while other environmental factors such as dietary habits are handled so-so at best.

Institutional mechanisms are in place to quickly monitor and adjust medications for many maladies like diabetes and heart disease. For example, there is a broad array of therapy options available if one is otherwise generally healthy but has heart disease. However, a patient with diabetes avoids many medications because they just don't tend to work well or have a tendency to exacerbate diabetic problems. This example is analogous to many other medical conditions. In other words, current patients are using the conventional wisdom learned from others who have taken the same medication earlier. This information is efficiently collected and passed on to physicians in a timely manner. But how much serious attention is given to the eating and nutritional habits of patients? Is the timing of taking a pill and what is also in the stomach a big enough factor to affect how well the medication might work? How big is this environmental factor compared to genetic variations among people? How often have these factors played an unknown role when the FDA has had to consider restricting the use of some medications?

Very aggressive action has often been taken for popular heart, arthritis and multiple sclerosis medication when rare, often sudden, severe adverse reactions leading to death have occurred – namely, the medication is often taken off the market. Now, to be taken *off* the market, the medication has to have been *on* the market. This means that it's already been through FDA approval based on extensive data showing a benefit and reasonable degree of safety. What happened? Assuming the benefit was real, then we are denying useful medication to many by cautiously protecting a few undeterminable unlucky ones. In other words, many lose so a few won't lose a lot. It would be nice to have better options rather than complete product removal when these problems arrive.

Current blood monitoring for liver intolerance of statins (a class of cholesterol-lowing medications) used for heart disease is a good example how these problems can be successfully addressed. Because there was a very broad patient base for statins, a huge profit incentive for finding a blood test existed – a good medical solution from good capitalistic motivation to find a fast, effective path to healthy patients and higher profits.

But how many rare reactions to medications are due to underlying genetics? Creating a large genetic data bank is a natural extension toward pinpointing who might be most likely at risk if genetics is a primary factor for some of these rare adverse reactions to medication. Rare reactions having mini-binomial clumpings around uncommon genetic markers will warn other similar geno-patients to shy away from some types of medication. This is merely an extension of our existing practice of acknowledging certain races have a predisposition to certain diseases. Genetics altering therapy options is a no-brainer. Effective tracking may prove to be very useful.

Question: Why isn't this already being done? Answer: It will be difficult to generate momentum from profit motivation because how can the pharma industry, as a whole, make any money doing this?[13] Profit margins are likely to go down for some companies if this concept is implemented. For example, a large genetic base monitoring system might very well determine that a huge segment of the population taking statins are not likely to gain any benefit. There is no profit incentive to do this as the number of patients taking medication is most likely to go down - not up. Companies don't like that. Just as diabetes restricts use of some medications - genetic information on the efficacy of medications will likely reduce overall medication use. Alternatively, this information would be useful for those few companies facing possible removal of their medications from the market because of rare adverse reactions that can be traced back to genetics.

Summary of Improving Restriction Options

Let's look at the recent example of the suspension of multiple sclerosis (MS) medication that appeared to be very effective at impeding the progression of MS. But due to its anti-inflammatory potency, deadly viruses occasionally enter into the brain. Rather than take this medication off the market, a more moderate future response might be:

---

[13] In fairness this is happening on a limited basis, but lagging, like the US Navy advancing airpower in the 1920s or the discouragement of smoking in France during the 1970s.

1) Acknowledge that the potency might often be high because of our efficacy threshold; therefore, lower dose 33-66 %.

2) Get genetic profile along with other pre-existing conditions.

3) Any candidate with a preexisting condition or genetic markers representing 2% or less of the general population that also existed in those who had an adverse reaction, would be considered a risky candidate for this medication.

This is likely a better option than removing medication that by-in-large worked well for such a catastrophic debilitating disease like MS.

This change in determining best therapy options assumes numerous mini binomial clumpings around many different genetic markers is, in fact, reality. The level of truth and success of this concept is unknown. Lacking a good profit opportunity suggests a publicly funded entity. But would we end up with a well run agency like NASA in the 60's or a self-serving boondoggle gravy train going to nowhere while searching for the right type of statistical facts to justify its further existence?

To recap:

1) Our statistical number-crunching exposes a shortcoming in current statistical methods. Efficacy statistical thresholds established to evaluate acute maladies may create *de facto* overdosing for slow degenerative diseases.

2) Genetic profiling and treating such information similar to pre-existing conditions may be a very useful tool for managing the bad reality that one binomial does *not* fit all. Many mini-binomials are a fact of life and fitting these mini-binomials into one big binomial assumption, while convenient, is a statistical sham.[14]

Not on the "Statistical Significance" Radar Screen

Our last look at health statistical parameters is one given less importance in the United States than in other industrialized countries like Germany and Japan. An American visiting a German or Japanese pharmacy will notice a striking difference in product lines and shelf space devoted to herbal and nutritional supplements. Americans looking for cultural oddities will notice how many Japanese now have a companion to their ever-ready camera – it's their pill box. Many Japanese are very tuned into the importance of the timing and order of consuming of food, nutritional supplements, and prescription medication.

American medicine pays little attention to this trend and tends to limit it to warning labels on prescription medication telling the patient to take

---

[14] See appendix B for further math on this dilemma.

"*My approach is nontraditional, but from a uniquely Western perspective.*"

the medication with or without certain pills and meals. Many American doctors start using words like "unfounded" or "quack" when discussing herbs and dietary supplements. Some papers might show a modest average difference between an herbal versus a control group, but seldom reach the level of statistical significance (P<.05). Hence no conclusion should be drawn. The exception in mainstream American medical thinking is the negative interactions that occur when taking prescription medication with herbal pills or foods like grapefruit. The current mainstream theme of American doctors: Herbal supplements can't help but might hurt. Hmm...pretty strong one-way corollary – no bias here?[15]

At the other extreme is how thoughtful the Japanese are in the consumption of everything. The meal starts with either protein or carbohydrates depending on one's later schedule. The concept of concentrated "good stuff" from foods and herbs put into pills is considered a no-brainer good idea in Japan. Which supplement pill to take depends on one's needs - like - does one want to be energized or relaxed? An American dining in Japan might see his counterpart take a few specific pills just before lunch, two in the middle of the same lunch and one last pill right before departing the table. These pills are taken out of a matrix box after a study of one's options. The whole behavior is reminiscent of viewing a fisherman studying his well-ordered tackle box and picking out the best lure.

So the Japanese have made a scientific expansion to nutritionals from the simple concept of taking coffee when waking up and not when trying to sleep. They take note of many subtle nuances in all foods and supplement pills they consume that may seldom show a statistical significance in medical research.

One interesting food paradox arose when it was learned grapefruit alters the potency of some medications. Grapefruit also happens to be very healthy for colons – particularly those with polyps. The change in grapefruit consumption was exactly the opposite in America than in Japan after the news broke. The American news media loves keeping consumers staying tuned through the commercials with worrisome teasers like, "Is grapefruit ruining your health by damaging your life-saving

---

[15] The FDA has an ever-increasing guideline for food/nutritional interactions with drugs. Grapefruit interacts with CYP enzymes and tends to decrease absorption of some antihistamines (e.g., Allegra 24 [fexofexahine]), while increasing the potency of other medications (e.g., some statins, buspirone and nisoldipine). A few of the many other nutritional items on the FDA guideline for interactions with drugs include a caution that St. John's Wort might adversely interact with oral contraceptives and heparin (the current "monster" in the guidelines), a blood-thinner affected by a host of good foods. (In other words, avoid eating healthy fruits and vegetables with heparin. Hmm ... a poster-child for more in-home monitoring of a therapy's effectiveness?)

medication? That story soon." This not-so-bottom-line information traveled very quickly, even though it was not told for another 22 minutes. The Japanese teaser story had a different theme: "A very healthy food means changes in some medications are needed." The Japanese inclination was to change the dosage of medications while American physicians were inclined to tell patients to stop eating healthy grapefruit. Grapefruit farmers saw their American demand evaporate while Japanese demand increased dramatically.

Same statistics - opposite reactions.

Are the Japanese being duped into believing unfounded benefits and quackery – or – is the old adage, "you are what you eat," more important and subtle than most Americans and American doctors think? The *right* answer is all relative to your past experiences, and which position best feeds one's ego.

Which society best addresses the shortcomings of statistically significant thresholds in our medical analysis? Who is improving their health and life expectancy faster, the Americans or the Japanese?[16]

If foods and supplements interact with drugs, then obviously powerful drugs interact with other drugs. More serious attention to multivariate evaluations will likely replace the concept of completion for single active ingredients in the not so near future. Unfortunately, the calamity of resistant bacteria was needed to highlight the benefit of multivariate evaluation for therapy. A simple concept can be used in a variety of ways to improve multi-medications and is the topic of Appendix C.

Twain's Summary for Today's Medical Statistics

Medical statistics can be improved, but will never trump good communication between patient and caregiver. Statistics will never surpass a good caregiver's intuition.

Factoring multivariate combinations of medication, diet and genetic interactions into a more complex post approval statistical data base will dramatically help patients and physicians. The tweaking of statistical efficacy methods is a pressing need for slow degenerative diseases.

Medicine is particularly vulnerable to the broad dilemma once described by Winston Churchill: "Capitalism is the unequal distribution of wealth. Communism is the equal distribution of misery." Where's the

---

16 World Health Organization (WHO) 2003 statistics: Life expectancy, U.S. 77, Japan 80 (But we now know how inconclusive this statistical fact is!)

proper balance between pooling everyone's resources versus individual liberty for medicine?

Arguments based on statistics justifying much more centralization of our medical system are always easy to make since they craft a logical tale with the knowledge of **hindsight**. But too much **centralization for foresight was the big lie** for half the world in the 20th century. Why should we perpetuate this lie for advancing our medicine now?

Statistics can easily mislead whether in politics, economics, medicine or quantum mechanics. The obvious paradoxes found in medical statistics should give one pause for an over-confident sense of certainty for the statistical reality called quantum physics.

We now turn to paradoxes surrounding unambiguous exactness. Or, maybe, vice-versa. Entering Einstein's world of physics.

Chapter 2:

Physics Before Einstein:
A Focus on Cem

It is best to learn a little history of basic physics to appreciate Relativity and why conservation of energy <u>and</u> momentum played such an important role. Conservation of energy <u>and</u> momentum, together, create a synergistic exactness yielding an infinite amount of knowledge that is not possible if one or the other cannot be assumed. This type of exactness is seldom found in medical and social statistics. We will view historical examples of using conservation of energy <u>and</u> momentum (Cem) to show why it is so useful in physics.

Early Goals and Dreams of Physics

Early physics concentrated on three practical areas: 1) instruments for measuring time; 2) wind and water mills, and other possible sources, used as a power source; and 3) instruments of war such as sailing strategy, cannon ball and arrow trajectory. Time instruments were often miniature mechanical gears analogous to the waterwheel. Waterwheel-powered flourmills were a stable of country manors near streams or rivers by the 15th century and optimists contemplated moveable waterwheel systems that could be a source of power anywhere. This was the concept of perpetual motion machines in its infancy. Newton's observation that gravity was a force throughout our world buoyed these optimists. They conceived clever portable watermill systems complete with moveable rivers or alternatively, unbalanced wheels that used leveraged weights on one side to perpetuate the lifting of un-leveraged weights on the other side of a wheel.

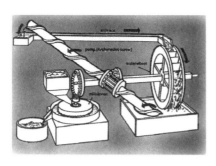 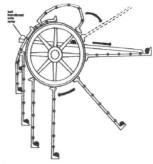

Pragmatists noted that Newton's most famous saying "Every action has an equal and opposite reaction" implies such portable waterwheels and forever rotating leveraged wheels will not create endless work. These perpetual motion machines rely on the gravitational force acting more on one side of a device than the other side forever.

The gravitational force on an object is described in algebraic terms by:

$$F = GM_em/R^2 = gm$$

F is the force of gravity, G is the gravitational constant, M_e is one mass (in this case earth), m is the other mass, and R is the distance from the center of one mass (M_e) to the center of the other mass (m). For an object (m) close to the surface of earth, R changes very little so $g = GM_e/R^2$ is considered a constant for practical estimates of the gravitational force near the earth's surface. The gravitational force described in the equation above, combined with the conservation of momentum, was used to explain the orbits of planets.

Many clever perpetual motion devices have been designed and sometimes made. But could any work? All human manipulations of $F = GM_em/R^2$ have failed to produce a working perpetual motion machine.

The concept of conservation of energy slowly developed as an aid in calculating the optimum gear mechanism and millstone size for maximum efficiency of the watermills. And as the Cem tool grew in applicability and acceptance, the dream of perpetual motion dimmed. By 1775 the French Academy of Sciences passed a resolution declaring that perpetual motion machines were impossible. How could energy be made where it hadn't already existed if energy is always conserved? Before 1850, the German von Helmholtz presented convincing arguments that conservation of energy implies the impossibility of perpetual motion machines. His circular logic also argued that the lack of a successful perpetual motion machine verified the absoluteness of the conservation of energy. The British and American patent offices made policies passively disclaiming the feasibility of perpetual motion devices as the evidence continued to pile up that they are likely impossible. These were a few of the many events that caused physics to develop the concept of absolute conservation of energy.

Meanwhile, the concept of a portable power-mill became a reality through boiling water rather than by the successful invention of a perpetual motion machine. Boiling water had run crude engine toys for over a thousand years. Amusing steam toys morphed into practical machines when James Watt invented the steam engine. Fine tuning steam engines to run efficiently was aided immensely by assuming conservation of energy. The world was also finding many other useful applications for Cem.

Physics Using Cem Conquers the World Militarily

England's military was also slowly relying more on the conservation of energy and momentum to help find solutions to complex problems. Military physics was largely the study of cannon fire during the early 1700s and 1800s. Physics helped Britain become the dominant world power for more than a century. For instance, accurate cannon fire was a major determining factor in the victory of the British Navy during the Napoleonic Wars. Cem was a key secret in getting this superior accuracy. Admiral Horatio Nelson's navy was very well versed in calculating how the different powder charges, cannon size and angle of the barrel to the horizon, and the mass of the actual cannon ball will determine almost exactly where the cannonball will land. The British Navy tediously mastered complex combinations of variables and so could easily sink an enemies' ship lacking their advanced mathematical skills. In two great navel battles, Admiral Nelson saved the British from having to learn proper French during the Napoleonic Wars. The French had more ships and the big advantage of land-based cannons at the mouth of the Nile River. The French had anchored their ships in a defensive half moon shape to bring as many guns to bear from shore and ship as possible. While the French had these many numerical advantages, Nelson had the numerical mastery of Cem. Nelson methodically proceeded to sink the French fleet with his longer range of accuracy.

The Trafalgar naval blow for Napoleon relied not on the British mastery of accuracy at a distance, but more on their knowledge of how to calculate quickly the next good trajectory based on the previous shot. Their superior knowledge of physics again proved decisive. So, the meticulous mastery of Cem enabled a small island nation to create an empire controlling all of the oceans and one-fifth of the world's land mass.

No wonder the embrace of Cem became a technological religion. The nation that best understood Cem practically controlled the world – literally.

The British relied on complicated tables or equations to calculate the momentum of the cannonball, how the energy of the gunpowder, and the force of gravity affected the cannonballs' trajectory. This can only be done by accurately quantifying and correlating all the major variables that determined where the cannonball will go. More importantly, equations had to describe how changing one component affected the other components. Cem filled in unknowns, as we will soon see in two simple examples.

Being the world's best – from waging war to making and transporting goods – depended on a superior knowledge of physics. It is no wonder that by the 1800s, Britain, Germany and France considered good physicists as national treasures. The same happened before World War II, when the flight of physicists menaced by Nazis before the war helped the United States and Britain to develop the atomic bomb. Moreover, the spoils of WW II included many German physicists who helped to develop the U.S. missile and space programs.

Some Basic High School Physics: A Bird's-Eye Overview

We will convert three sentences describing major components of cannon physics to algebra equations. Basically, one equation represents a straight line and the two other equations are curves. Some intersections of these lines and curves prophetically predict exactly where a cannonball goes. This will give us the tools and observations to see the importance of Cem in two very simple examples.

1) The momentum of an event is conserved in a linear fashion with respect to velocity. It is the mass of the cannonball times its velocity (speed)

$$\text{Momentum} = mv \rightarrow (\text{mass}) \times (\text{distance})/(\text{time}) \text{ in units}$$

2) The energy of a cannon ball is not linear, but squared to its velocity.

$$\text{Energy} = E = \tfrac{1}{2} mv^2 \rightarrow \tfrac{1}{2} (\text{mass}) \times (\text{distance})^2/(\text{time})^2 \text{ in units}$$

where E = energy of the gunpowder, m is the mass of the moving object(s) and v is the velocity. Hence the momentum of a cannon ball is,

Mom= M_c x V_c, while its energy is E= ½(M_cxV_c^2), where M_c is mass of cannon ball and V_c is the velocity of the cannon ball. The cannon's recoil is then half of the gunpowder's energy and the other half of the energy sends the cannonball on its way. (This is where Newton's famous saying, "Every action has an equal and opposite reaction." was so useful.)

3) Gravitational force is also not linear. It is often more complicated than the energy equation - yet for other problems – it can be simplified, or eliminated altogether.

Gravity's Force = GM_em/R^2 = gm (close to earth's surface) $\rightarrow$
Even more units! [17]

Gravitational equations not only help to predict cannon fire, but they can also describe the beautiful elegance of water fountains.

Luckily, our examples of Cem will be so simple that we won't even use a gravitational equation.

One can explain much of classical physics using the tools of mathematics, the three equations of momentum, kinetic energy, and gravity's force –

Momentum = mv, Energy = ½ mv^2,
Gravity's Force = GM_em/R^2,

<u>AND</u> assuming absolute conservation of energy <u>and</u> momentum.

Correlating linear momentum to a non-linear energy to a different non-linear gravitational acceleration to give exact results is not easy without computers.

The Beauty of Cem

The importance of Conservation of Energy <u>*and*</u> Momentum (Cem) in physics is quickly learned by looking at cannonball physics applied to billiards.

A good pool player has an intuitive sense about momentum and energy. This instinct of pool play is made possible by the fact that the conservation of <u>both</u> energy <u>and</u> momentum (Cem) mandates a precise

---

[17] The mass is typically in kilograms (kg,) the distance in meters (m), the time in seconds (sec), and the force units are called newtons (nt) in the three classical equations. G is the gravitational constant (its units are; nt-m^2/kg$_2$), M_e is the mass of the earth (units in kg), m is the mass of the object(kg) , and R is the distance to the center of each mass.

outcome. Einstein argued that if we know exactly how a player hit the cue ball with respect to the speed, direction and spin of the cue ball; and if we know the exact friction and all other external forces; then we could predict <u>exactly</u> what would happen to all the pool balls on the table - no matter how many balls are on an infinitely large pool table. That claim is still hotly debated by physicists.

Two of the simplest ways pool balls can strike each other will give an appreciation for the beauty of Conservation of Energy <u>and</u> Momentum.

Example 1

The first example is the cue ball hitting another ball dead center so the hit ball is sent traveling in the exact same direction as the cue ball was traveling. We will say the cue ball is traveling at 10 miles per hour (MPH) and all pool balls weigh exactly 1 kilogram (Kg) to keep the arithmetic simple. How does conservation of energy and momentum predict an exact outcome?

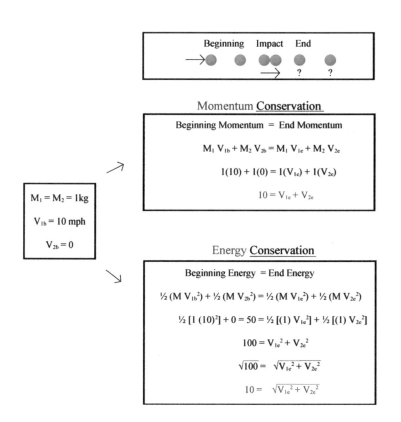

Beginning   Impact   End

$M_1 = M_2 = 1\text{kg}$

$V_{1b} = 10 \text{ mph}$

$V_{2b} = 0$

Momentum <u>Conservation</u>

Beginning Momentum = End Momentum

$$M_1 V_{1b} + M_2 V_{2b} = M_1 V_{1e} + M_2 V_{2e}$$

$$1(10) + 1(0) = 1(V_{1e}) + 1(V_{2e})$$

$$10 = V_{1e} + V_{2e}$$

Energy <u>Conservation</u>

Beginning Energy = End Energy

$$\tfrac{1}{2}(M\,V_{1b}^2) + \tfrac{1}{2}(M\,V_{2b}^2) = \tfrac{1}{2}(M\,V_{1e}^2) + \tfrac{1}{2}(M\,V_{2e}^2)$$

$$\tfrac{1}{2}[1\,(10)^2] + 0 = 50 = \tfrac{1}{2}[(1)\,V_{1e}^2] + \tfrac{1}{2}[(1)\,V_{2e}^2]$$

$$100 = V_{1e}^2 + V_{2e}^2$$

$$\sqrt{100} = \sqrt{V_{1e}^2 + V_{2e}^2}$$

$$10 = \sqrt{V_{1e}^2 + V_{2e}^2}$$

Where; V_1 = cue ball, V_2 = ball hit, V_b = beginning velocity, V_e = velocity after collision.

Momentum is defined as mass times its speed or velocity (MxV). Many experiments have proven the energy of the cue ball to be ½ MxV². Where M is the mass (weight) of a pool ball and V is its velocity for both equations.

Algebraic substitution, incorporating one equation into the other, will show that 10 & 0 are the only possible solutions to both equations. So the cue ball comes to a complete stop and the ball hit leaves with the exact same velocity as the cue ball. This is the only possible outcome if Cem holds true. The following graph is an intuitive aid to seeing how Cem works so well in this very simple problem.

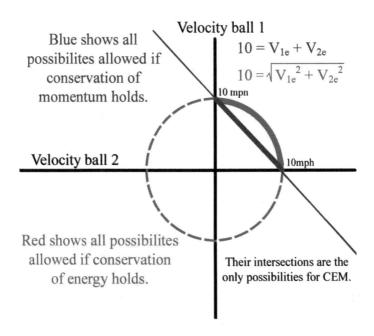

Any point (coordinates) inside the red circle would mean energy was mysteriously lost and outside the red circle meant energy would mysteriously be gained. These two possibilities are impossible because of conservation of energy. But numbers right on the solid circle line are all possible if conservation of energy is maintained. The same is true for the blue line where one side is an increase in momentum and the other side of the line is a decrease of momentum. Their intersection is like the convergence of azimuths in old naval navigational positioning techniques. Pool players know from experience that these math results are correct.[18]

Example 2

(The non-math reader should just skip to the underlined sentences on page 59.)

We will have the cue ball hit exactly and equally between 2 balls that are touching one another. The cue ball will come to a complete stop upon impact. Can we calculate exactly the speed and direction of the 2 balls hit? We will use the same mass for the balls and the same starting velocity of 10 MPH as in Example 1. The fact that the balls are symmetrical and equal and the cue ball stops, means we can predict velocities without even using Cem. Then we can use Cem to check our answers.

We will look at conservation of momentum first:

$$\text{Beginning Momentum} = \text{End Momentum}$$

$$M_1V_{1b} + M_2V_{2b} + M_3V_{3b} = M_1V_{1e} + M_2V_{2e} + M_3V_{3e}$$

$$\text{Since } M_1 = M_2 = M_3 = 1Kg \ \& \ V_{1b} = 10$$

$$1(10) + 1(0) + 1(0) = 1(0) + (1)V_{2e} + (1)V_{3e}$$

Now since the hit is exactly between the two balls they must depart with the same speed. So, $V_{2e} = V_{3e}$

$$10 = V_{2e} + V_{3e} = 2\,V_{2e}$$

$$5 = V_{2e} = V_{3e}$$

---

[18] Our examples have assumed no external forces are affecting the balls to keep the math simple. We also assumed the balls skid across the pool table and are not rolling. This simplifies our equations a lot because a ball having a roll or spin contains an internal "rotating momentum and energy". Any change of this "rotating momentum and energy" from one ball to another would need to be added into our calculations. Spinning balls change reasonably simple equations into complex equations – as any good pool player already knows.

Now we will now look at conservation of energy:

$$Beginning\ Energy = End\ Energy$$
$$(½)(M_1[V_{1b}^2] + M_2[V_{2b}^2] + M_3[V_{3b}^2]) =$$
$$(½)(M_1[V_{1e}^2] + M_2[V_{2e}^2] + M_3[V_{3e}^2])$$

Again since all $M_s = 1$, and $V_{1b}=10$, and $V_{2b} = V_{3b} =0$
so the above equation becomes;

$$Beginning\ Energy = End\ Energy$$
$$(½)(10^2 + 0 + 0) = 50 = (½)(1[0^2] +1[V_{2e}^2] +1[V_{3e}^2])$$
$$so,\qquad 50 = (½)([V_{2e}^2] + [V_{3e}^2])$$
$$but\ V_{2e} = V_{3e}$$
$$so\ (50)^{1/2} = V_{2e}$$

Something is very wrong: 5 mph for momentum does not equal a number slightly larger than 7 mph for energy. Shouldn't these two numbers be equal?

What happened? Cem will show us our faulty logic.

In this case, the conservation of momentum is the momentum of the two balls traveling in the exact same parallel direction as the first ball. The balls are also traveling in equal opposite directions perpendicular to the direction of the original ball. Since both balls are moving away from the direction of the cue ball equally, their net momentum other than the cue ball's original direction is zero. Let's assume the 5 MPH represents the speed the balls are going in the same direction as the cue ball. But we have to add the perpendicular velocity of these balls to get a total velocity. Let's check this thinking by using the Pythagorean discovery in 88 AD that the square of the hypotenuse is equal to the sum of the squared sides of right triangles. Will this work for Cem?

This gives us for energy; $(50)^{1/2} = (5^2+x^2)^{1/2}$ where x is the perpendicular velocity of each ball from the original cue ball's direction of travel. Solving this equation gives us x=5. (Again, the perpendicular velocity of 5 MPH for each ball is in the exact opposite of the other ball so the net change of side momentum is zero.) From this we also know that the balls leave at exactly a 45 degree angle from the direction of the pool ball and the two pool balls will be traveling exactly perpendicular to each other at $(50)^{1/2}$ miles per hour. <u>Cem was essential for telling us the exact speed and direction of each ball without the need to actually hit the balls!</u> We have certainty with Cem that this will be the exact outcome for the pool balls under these conditions – Because this is the only possibility. <u>Cem eliminated chaos</u>. This exact outcome has also been seen by many experienced pool players to be true. Cem also told us our final method of adding perpendicular velocities in this example was correct.

The use of Cem is even more powerful and useful when studying atomic particles where our eyes cannot easily check our math. Physicists must use much more complicated views of momentum and energy than in these two very simple examples. Cem is indispensable – our seasoned physicists can feel defenseless against uncertainty without it!

Our initial logic in our second example was both right and wrong. We started with two fairly reasonable but conflicting conclusions: if we just calculated velocity from $\frac{1}{2}mv^2$ and converted those speeds to momentum, then our new total momentum is around 14; while our other answer using Cem maintains the momentum at 10. This is a little like the contradictions in our aspirin and Tylenol study having two opposite – yet, both fairly logical conclusions in Twain's World of statistics. But unlike the uncertainty that remains in statistics, our conventional wisdom of Cem explains the conflict and tells us which answer is useful...and which is wrong! Cem is powerful stuff.

Another Example of Sound Logic Gone Wrong

The difficulty of physics is seldom appreciated unless one is given an unfamiliar problem to solve. Seemingly logical reasoning is often wrong in classical physics. Choosing the correct options in physics is not always obvious. The old saying, "What goes up, <u>must</u> come down," is a good example of logical but wrong reasoning. This saying applied to projectiles leaving a cannon muzzle and was a popular truth in the 19th century...except for a few eccentric mathematicians and blimp enthusiasts. "What goes up, must come down" fit all practical examples of cannon projectile theory of the 1800's. A fairly convincing proof can show this saying is all encompassing by using basic gravitational and velocity equations. The logic was: an object is always being attracted to earth causing a decrease in any upward velocity. Hence, its speed leaving the earth is always slowing no matter how fast the projectile's initial velocity. Therefore the projectile will eventually stop leaving the earth. It will then proceed to fall back to earth because earth's gravitational attraction exists no matter how far an object is away from earth.

Obviously, that's not how the universe works as the departure of the first space probe from our solar system demonstrated. The flaw was assuming that an object always slowing will eventually come to a stop. Those early physicists who embraced this obvious conventional wisdom did not consider that even if you add an infinite number of progressively smaller numbers you may never reach a certain specific number or threshold – like continually dividing in half a number will never get you to exactly zero. A specific example is adding together a series of infinite numbers that progressively become smaller but never reach zero, but interestingly, their combined total also will never exceed the number 3.142.

A simple mind analogy of this fact is to draw a line in beach sand and then stand an arbitrary distance from the line. Now move ½ the distance to the line, stop and yell, "All logical assumptions describe reality." Now move ½ the distance to the line. Again stop and yell the statement again. Do this over and over. When will you reach the line? Will you ever travel beyond the line while doing this? Will you ever stop getting closer to the line? The line can be thought of as the point when the projectile comes to a stop – you're always getting closer but never reach – as in coming to a complete stop – and beyond the line is the projectile falling back to earth.

The escape velocity for earth can be calculated using calculus. It is about 7.0 miles per second.[19] If the projectile is traveling less than 7.0 miles per second, then our line in the beach sand can be used as a weak analogy to describe how the projectile never leaves the earth's clutches and hence will fall back to earth. Hence this half-way to the line mind game might be useful in explaining escape velocity concept to someone unaware of the debate, but useless in changing the opinion of many already convinced that "what goes up, must come down" is correct. In fact, they can easily believe that this argument helps rather than counters their line of reasoning!

How many of our current scientific beliefs are flawed by logical, but fatally wrong assumptions – like the antiquated truism, "What goes up, *must* come down"? Much of Twain's and Einstein's genius was questioning concepts based on sound logic.

Physics Just Before Einstein's Discovery

It was an exciting time for physics in the late 1800's. The debate about, "What goes up must come down" was just one of many lively questions. Physicists Hendrik Lorentz and George FitzGerald were making

---

[19] See appendix D for the math.

observations showing that pool ball and cannon ball physics was seriously contradicted by the bizarre behavior of high-speed particles – Cem appeared to be very flawed. Meanwhile, Marie & Pierre Curie were finding a seemingly endless supply of energy from pitchblend (radium).

A revival of possible perpetual motion machines was contemplated by both optimists and charlatans as this mounting strangeness in physics was growing. Yet, Maxwell's famous four equations appeared to accurately describe all of electricity by 1890. These four equations made incredibly accurate predictions when using Cem.

But, there was clearly a dilemma. Why were the Lorentz and FitzGerald's results so far off from predictions of cannon ball physics? Einstein revolutionized high speed physics in 1905 by answering this question. Also, why were the Curies finding seemingly endless sources of energy from pitchblende? Einstein helped answer this question in 1916. Einstein incorporated mass as part of the absoluteness of the conservation of energy and momentum (Cem) into his original 1905 paper to derive his theory of General Relativity. General Relativity was solid groundwork for explaining the strange energy from radium. A new definition of Cem was the foundation for General Relativity, and General Relativity reaffirmed the usefulness of Cem.

Einstein helped bridge a gap between the physics of pool/cannon balls and the Maxwell equations of electricity and magnetism. His theme can be amusingly named, "Don't put your faith in the absoluteness of your clock or tape measures, but rather in mathematics and the absoluteness of conservation of energy and momentum." General Relativity is used to make predictions regarding the smallest particles to the fate of the universe.

"I think you should be more explicit here in step two."

Chapter 3

Relativity Made Easy: Deriving the Backbone that Supports All the Other Stuff

In this chapter, we will see how time is slowed for fast moving objects and how Einstein derived his eloquent equation: $E=mc^2$.

As mentioned in the introduction, Einstein literally daydreamed his way to answering questions that had pestered some very bright and hard working physicists for more than a decade.

Even today, a popular topic of discussion amongst physicists is the fact that Einstein's equations of 1905 had already been derived from experimental results by Lorentz. So, you had one scientist, Lorentz, who spent years in a lab carefully gathering data and tediously deriving a formula. Yet another amateur scientist, Einstein, who in his fantasy world of mathematics conjured up and explained why the equations occurred! He used nothing more than pen, paper and his imagination.

Einstein was an indifferent student, perhaps even "learning disabled" in his dreamy world. Would Einstein have lost some of his natural abilities if as a child he had been medicated to remain focused on his lessons, instead of practicing his daydreaming talent? Is there a downside to medicating boys and girls who think like Einstein so they will passively listen to the teacher and not daydream their lives away? But we digress... as daydreamers often do.

So, here are the famous discoveries of a habitual daydreamer:

The underlying foundation to Einstein's discovery of Relativity is an unusual characteristic in light that also applies to fast moving objects. We will call this "the trick"!

The Trick!

It is easier to know the secret of something magical if you understand when and where "the trick" occurs. The secret to understanding the "magic/trick" of Relativity is very simple. However, it is in subtle contradiction to our everyday experiences. The physicist is the magician and an unusual characteristic first discovered in light is the trick in Relativity. Below is a simple unused example of the "fact/trick" to understanding why Relativity exists:

A Simple Fact that Makes the Theory of Relativity Possible

We are used to taking for granted that throwing a snowball at a mailbox will travel at different speeds (velocities) if thrown by a person standing nearby versus a person approaching the mailbox in a pickup. For example:

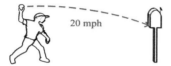 

If the person throws a snowball hard enough that it leaves his hand at a speed of 20 MPH, then the snowball will hit the mailbox at roughly 20 MPH.

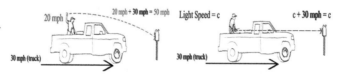

But if that person is standing in the back of a pickup heading directly for the mailbox at 30 MPH then the snowball will travel at the speed of both the throw and the pickup. It will hit the mailbox at about 50 MPH.

On the other hand if the pickup is driving away from the mailbox at 30 MPH, a 20 MPH toss by the passenger will never reach the mailbox. Instead it will travel at 10 MPH away from the mailbox (i.e., will be moving away from the mailbox at 10 MPH).

> *Before you yawn and lose interest…**Light does NOT act this way!** Instead of throwing a snowball as in the above examples, shine a flashlight on the mailbox. The speed (velocity) of the light (denoted as C) hitting the mailbox is the exact same in all three examples! The speed of light is 186,282 miles per second=C. The light traveling from the pickup heading towards the mailbox is 186,282 miles per second, not 186,282 miles per second plus the 30MPH speed of the truck. Finally, the light traveling to the mailbox when the pickup moving away from the mailbox at 30 MPH is 186,282 miles per second not 186,282 miles per second minus the speed of the truck. Even if the pickup could be souped-up to travel at 10,000 miles per second, many experiments have shown that the speed of light would be precisely the same —186,282 miles per second—for each of the above examples for an observer standing near the mailbox! This fact creates paradoxes for classical physics that Einstein resolved.*

How this fact appears to make magic in the right circumstances will now be demonstrated in two examples. In the first example, we will use junior high math skills to see how time is slowed for fast moving objects. This is one of the famous curiosities about Relativity that is popular in many science fiction movies. In Example 2, we will see how our first example, our knowledge about momentum from electrical experiments, and ordinary high school math skills gave Einstein the tools to develop his most famous equation, $E=mc^2$.

Example 1: Why Time Can Be Warped

Suppose two imaginary space vagabonds decide to travel to a distant star together in parallel moving crafts, like a pair of motorcyclists traveling on a country highway. They travel to this star at a speed approaching the speed of light. From time to time they might exchange messages by flashing a light directly from one craft to another. This light signal would appear to these travelers as:

The signal travels the distance $C \times T'$ to get from one craft to the other craft. C is the speed of light and T' is the time clocked on the vagabonds' watches for the light to reach the other craft a distance d_1 apart.

Now imagine an outer-space hermit busy going nowhere. He is just enjoying his daydreams in the solitude of outer space when these spacecrafts zoom past him while sending their laser message. The light message would have looked different to the hermit than to the travelers. His supernatural eyes would see:

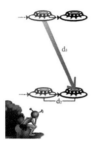

Crafts traveling at speed V for time T until the signal reaches the other craft will travel distance d_2

Now, if we superimpose both the observations of both the hermit and the vagabonds we have:

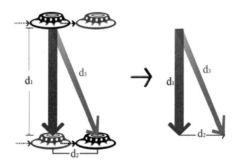

We note we have a problem with d_3 being longer than d_1. Since <u>light **is** **always** traveling the **same speed**</u> for both d_3 and d_1 – <u>We have a paradox since they are not the same length</u> – our "trick" has just occurred. We have a mathematical contradiction...we can give up on this obvious nonsense – or follow Einstein's path of wilderness exploring in a paradox world of contradiction. We'll quantify this paradox and see where it leads us.

This right triangle of distances is determined by the speed and time involved in each side. In 88 AD, the Pythagorean cult discovered that the square of the hypotenuse is equal to the sum of the squared sides of right triangles. This ancient knowledge gives us:

$d_3{}^2 = d_2{}^2 + d_1{}^2$ where these distances are calculated by speed x time.

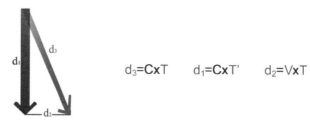

$$d_3 = C x T \qquad d_1 = C x T' \qquad d_2 = V x T$$

We end up with the equation:

$$(CxT)^2 = (CxT')^2 + (VxT)^2 \text{ where:}$$

T is time observed by the hermit; T' is time observed by vagabonds, C is the speed of light, and V is the velocity of the spacecrafts as observed by the hermit.

How does the equation $(CxT)^2 = (CxT')^2 + (VxT)^2$ hold up at first glance? First, assume V is much smaller than C (i.e., V<<<C). Then $(VxT)^2 \to 0$ compared to d_1 and d_3. Hence $(CxT) \approx (CxT')$. This makes sense.

Now we'll look at what we've been told since junior high school, namely that Relativity causes time to slow within fast moving objects as compared to slow moving objects. Let's check:

As $V \to C$ then $(VxT)^2 \to (CxT)^2$ therefore $(CxT')^2$ must shrink. <u>Since C is constant from all observations,</u> T' $\to 0$ as V→C. Therefore time is slower for the vagabonds. The closer the vagabonds are traveling to the speed of light as viewed by the hermit, the slower their clocks run compared to the hermit's clock. This holds up to what we have been taught when first introduced to the concepts of Relativity.

Let's optimistically tackle momentum by looking at the return trip of our vagabonds before examining the accuracy of

$$(CxT)^2 = (CxT')^2 + (VxT)^2 \text{ further.}$$

Example 2: How to Derive E=mc²

We will use observations about momentum on the return trip of the vagabonds to derive E=mc².

As we know from our pool table examples, momentum is the mass of an object times its speed. A one-pound snowball traveling 20 mph has the same momentum as a two-pound snowball traveling 10 mph. Everyone has seen how pool players instinctively understand how glancing the cue ball off another ball at a precise speed and angle will impart a predictable momentum and direction, sending the ball to a particular pocket. Physicists have studied momentum and angled interactions not just on pool tables, but also with electrons and protons in the laboratory. We will use what they have learned in this example.

Suppose our two space vagabonds have finished exploring the star system they set out to visit and are ready to go back home. However, they both want a little more room in their crafts on their return trip. They decide to make a barge that will carry their gear and souvenirs back with them. They place the barge perfectly between their two crafts as they travel back home. Once they and their barge reach their cruising velocity, our vagabonds enjoy the comfort of their roomier space. They are heading in the opposite direction than in example 1. So they again pass by the space hermit. As they pass the hermit, both crafts simultaneously emit equal laser pulses directly at the center of the barge as shown in the diagram below. Let's make educated guesses at what the vagabonds and hermit observe.

Above is what the vagabonds see; below is the hermit's view.

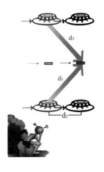

The vagabonds see the laser pulses counteracting each other. So, no change to the barge's forward momentum occurs.

However, the hermit sees the crafts and barge move forward as the pulses travel to strike the barge at an angle, thereby increasing the forward momentum of the barge! Can we quantify this increase in forward momentum observed by the hermit?

Again we'll superimpose the different (relative) observations of the vagabonds and the hermit.

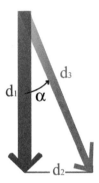

The hermit's view of forward momentum change can be estimated by looking at one of the identical triangles more closely.

We know from basic trigonometry that an angle α between d_1 and d_3 has a $\sin\alpha \equiv d_2/d_3$

Since $d_2 = V \times T$ and $d_3 = C \times T$, this means $\sin\alpha = V/C$

What do we know from classical and electromagnetic physics that might help us?

First is how energy, (denoted as E in algebra equations), can change momentum. Photons impart a little push of energy when they strike. The energy (E) for this has been observed in *many* different electromagnetic experiments to give a corresponding momentum change of E/C to a given mass. Since there were two equal pulses of energy from opposite sides, we will define ½ E as the energy from each pulse for the equations in this example.

Second, *numerous* observations have shown an angled addition along an axis will result in $\sin\alpha$ fraction of the total force or energy and a $\sin\alpha$ momentum change for that axis. Hence the forward energy from each side is adding energy of ½E($\sin\alpha$) if one assumes complete transfer of energy to momentum. Each side should therefore give a forward momentum of ½(E/C)(sinα). This is based on observations of *many* experiments.

Third, momentum is mass times velocity. This is conveniently similar to that of distance which is time multiplied by velocity, used in our Example 1. So our initial algebra equation is somewhat familiar to us. For example, the distance traveled by a space craft in Example 1 was (VxT) and the momentum of the space craft is (V x M_c) where M_c equals mass of the space craft. The momentum of the barge is (V x M_b) where V is the speed of the barge and M_b is the mass of the barge.

We now have enough information to put together an algebra equation. Let's form an equation and play with it:

Initial Momentum + momentum from pulses = New Momentum as seen by the hermit. Using algebra symbols, we have:

$$M_bxV + ½ E/C(sinα) + ½ (E/C)(sinα) = M_b'x\ V'$$

where Mb'xV' is the new momentum. So,

$$M_bxV + (E/C)(sinα) = M_b'xV'$$

but we know sinα = V/C. So,

$$M_bxV + (E/C)xV/C = M_b'xV'$$

(a) $M_bxV + E/C^2xV = M_b'xV'$

Now lets define M_b x V as momentum before the pulse and M_b' x V' as momentum after the pulse as viewed by the vagabonds.

Now the trick: How do the vagabonds see the momentum and is there conservation of momentum?

In other words, does M_b' x V' = M_b' x V'? We will assume Cem. We know from the diagram that M_b' x V'= M_b x V. V' & V' is the forward speed after the pulse. The initial velocity of the vagabonds as compared to the hermit's speed, V, and the initial velocity viewed by the hermit were equal. So V=V. (Remember, Blue is vagabond's view, Red for hermit.) But the vagabonds observe no difference, so V = V= V'.

Let's now look at the difference between the hermit's view of momentum after pulse, ($M_b'xV'$), and the vagabonds' view of momentum after the pulse, ($M_b'x$ V'). (i.e., ($M_b'xV'$)-($M_b'x$ V') =?)

Let's assume (Cem & the trick!) $M_b'xV' = M_b'$ x V'. This is true if Cem is absolute regardless of relative observation.

Since $V=V'$: $M_b'xV'= M_b'$ x V substituting the other side of equation (a) for $M_b'xV'$, we have;

$$M_bxV + E/C^2xV= M_b' \text{ x } V.$$

We can cancel V and put the masses on one side.

We now have $E/C^2= M_b' - M_b$. But what does this mean?

We have options:

Option 1: We can assume $M_b = M_b'$. So $E/C^2=0$. What might this mean? We have a flawed contradictory equation having no basis in reality; or energy is nil (another contradiction); or energy is coming from nil which is still another contradiction if conservation of energy is all encompassing.

Option 2: Assume M_b & M_b' are not equal and let m equal the difference between the two masses. In other words, m $=(M_b - M_b')$.

$$\text{Then } E/C^2= m.$$

Rejecting Option 1 for Option 2 gives us $E/C^2= m$.

This can also be written as $E=mc^2$

(Note: Our Example 2 is very similar to presentations made by both Einstein, in 1950, and later by Robert Resnick, in 1972, to derive $E=mc^2$.)

The next chapter will look more closely at equations
$$(CxT)^2 = (CxT')^2 + (VxT)^2 \text{ and } E=mc^2.$$
The math surrounding these two very eloquent equations gets complicated very quickly.

Chapter 4

So You Want to Know More about Relativity: You Got It!

We came up with $(CxT)^2 = (CxT')^2 + (VxT)^2$ in a rather straight-forward fashion. However, we made several reasonable but fairly speculative assumptions before arriving at E=mc^2. Ironically, Einstein had rejected $(CxT)^2 = (CxT')^2 + (VxT)^2$ as incomplete before 1905. But E=mc^2 was published in 1916. Einstein has been quoted as saying he was playing with Relativity at the age of 16. There were many popular algebra sailing games for youngsters in Germany one hundred years ago. Einstein likely used this algebra logic found in his love of sailing to create Relativity if his original path to discovery was similar to our thought experiments. The logic of using Example 1 to arrive at Example 2 is fairly straightforward for a naturally talented mathematician. Therefore one wonders if Einstein had arrived at E=mc^2 only hours after first deriving $(CxT)^2 = (CxT')^2 + (VxT)^2$, and waited more than ten years to emphasize the importance of E=mc^2 to the world. If so, why the delay? His General Theory of Relativity, based on conservation of energy and momentum, is accepted as an absolute truth by most physicists today.

The problem with $(CxT)^2 = (CxT')^2 + (VxT)^2$ is that it does not tell the whole story. We can rewrite;

$$(CxT)^2 = (CxT')^2 + (VxT)^2 \quad \text{as} \quad T' = T(1 - V^2/C^2)^{1/2}$$

using simple algebra.

As it turns out, the flaw in our assumption is that only time changes to make our Pythagorean Theorem "fit." The other component that made up this triangle is units of length. Modern physicists assume our triangle will also change length units along with time units in an equal proportion. This change in length has been seen in experiments for well over 100 years.

So not only do the vagabonds' clocks go out of sync with the hermit's clock, but their rulers change too!

Our original prediction of time change in Example 1 for the vagabonds was:

$$(CxT)^2 = (CxT`)^2 + (VxT)^2 \rightarrow T` = T(1- V^2/C^2)^{1/2}$$

Changing distance with time is algebraically more complicated than Example 1.

This yields the same $T`$.

$$T` = (1- V^2/C^2)^{-1/2} (T- V/C^2 \ x \ ) = T(1- V^2/C^2)^{1/2}$$

And also an additional corresponding distance measurement, x and $x`$ change along the vagabonds' direction of travel of;

$$x` = (1- V^2/C^2)^{-1/2} (x-VT)$$

We needed the correct assumption that space is changing together with time! It makes sense that if we choose to make *time* warpable, then *space* should also be considered as possibly warpable when trying to make the right Pythagorean triangle. However, our flaw in Example 1 did not prevent us from moving forward and deriving E=mc2 in Example 2! Our corrected Example 1 is the basis for **The Special Theory of Relativity.** Example 2 in our last chapter is the foundation for **The General Theory of Relativity**.

We will bypass the typical "first stop" look at paradoxes examined in most relativity books – the varying complexities of time and space depending on an observer's reference – and instead go directly to intrigues with Cem.[20]

Complications with General Relativity can be appreciated by continuing the vagabonds' journey to its end in Example 2. The hermit sees the barge pulling ahead of the vagabonds' crafts, yet the vagabonds observe no increase of forward momentum of the barge – yet Cem has been mandated in our mathematical assumptions! Hmmmm. How much

---

[20] Relativity is full of both apparent paradox and extreme detail. Appendix A, Twain's first paradox, is a good example of how a very simple paradox can create a complicated multi-dimensional-math landscape quickly.

energy would be needed to stop the barge as viewed by the vagabonds versus the hermit? [21] Does the barge come to rest between the crafts or ahead of the crafts if all are slowed to a rest simultaneously? The math becomes very involved when describing probable answers. This Example 2 dealt with a change of velocity morphing masses – a constant issue when gravity's acceleration is bumping into the light speed barrier. One can easily appreciate how the mathematics for gravity's morphing of time, space and mass/energy becomes quite daunting within "acceleration – the C barrier – equivalence boundaries."

Questionable Assumptions from the Start?

Talented physicists have argued that while the answer to Example 2 appears to be right; the reasoning is flawed as the focus should be on the lack of effect energy has on fast moving objects.[22] Suffice to say that we'll leave the bulk of these discussions to more technical books.

Right or wrong, Example 2 is an entry into General Relativity offered by Einstein. General Relativity assumes: the speed of light, C, is the top speed of everything; a correlation of mass being a form of energy has been established (i.e., $E=mc^2$); there exists no preferred reference (or relative) position, extending to a concept of equivalence for gravity and electromagnetic forces on masses; and conservation of energy and momentum (Cem) always holds true. Volumes of equations and massive books have relied on these assumptions, which are reasonable expectations when describing, at the very least, a huge subset – if not *all* of reality. These are the practical boundaries for Relativity's current conventional wisdom.

Most Relativity equations rely heavily on the absoluteness of Cem. And so as with Nelson's cannonballs, this absolute faith in Cem appears to yield great results for modern physics and for Einstein.

A brief historical look at Relativity in its infancy is interesting and helps explain the reason for the focus and direction of current research in Relativity. Minkowski's four dimensional geometric ideas aided in Relativity's early conceptual development. Many physicists, including Einstein, became fixated on the curvature aspects of Relativity and how

---

[21] The popular concept in sci-fi of parallel universes comes from these results originating from mandating Cem and $V = V = V'$.

[22] Mathematical 'funhouse' semi-enantiomers can sometimes be distilled to equivalent equations. An equation from experiments which Einstein was well aware of prior to 1916 is $E = mc^2/(1-(v^2/c^2))^{1/2}$. As $v \to 0$, then $E = mc^2/(1-(v^2/c^2))^{1/2} \to E = mc^2$. This is the famous rest mass equation.

that may alter the Pythagorean and other assumptions. The famous Bernard Riemann, who did much to clarify integral calculus in the 1800s, also worked on curvatures and developed curvature tensors, which contained 20 distinct components. Einstein adapted some of these components for advancing his ideas on Relativity. Theodor Kaluza, Nathan Rosen and an entourage of physicists worked on and off with Einstein in their task of trying to find mathematical simplicity from evermore complicated dilemmas. Many aspects of theoretical General Relativity are math analogies to the philosophical question, "Which came first the chicken or the egg?"

Philosophy of Determinism versus Uncertainty: A Wrong Turn or Digression?

Einstein strongly believed <u>exact</u> predictions could be made in all studies of physics. He optimistically thought solutions to complicated questions were right over the horizon of our knowledge. Physicists who are inclined to believe Einstein's contention that exact predictions are possible for almost all events in physics are called Determinists. Here is where Einstein and many physicists part company. Most don't believe this noble goal is always possible – particularly for quantum mechanics. By the 1930s, Werner Heisenberg was presenting compelling logic that we will never be able to precisely measure both position and velocity of small high-speed objects. If we can't measure their performance exactly, how can we possibly predict their future behavior exactly? At the same time, the mathematician Kurt Godel was getting attention by elaborating mathematically on analogous equations to sentences like, "This statement is wrong," or "Our only certainty is uncertainty." Godel was bringing uncertainty and the lack of our abilities to find answers to a new plateau compared to Heisenberg's arguments of uncertainty.[23]

Physic's vector equivalents to our sex discrimination statistical paradox were addressed specifically by neither Godel nor Heisenberg. We know the answer to this vector dilemma in physics for small number interactions happens to be quite contrary to the most pragmatic

---

[23] Godel's insight into uncertainty's implications is broad and far reaching. Godel's work on describing uncertainty can also be seen in specific successes of others. For example, Cicero's solution to a Godel-like problem was described two thousand years ago in his book, *The Commonwealth*. His theme tackled an uncertainty within humanity that caused his execution. His post mortem triumph over this human flaw was how his analysis was used as a blueprint for crafting America's remarkably successful system of government. So much of modern success can be attributed directly back to Cicero and how he proposed to deal with political/governing uncertainty. Yet, *The Commonwealth* is a masterpiece largely unknown by Americans. Einstein and Cicero dealt with different paradoxes but, whether liked or not, both confronted uncertainties of interest to Godel.

conclusions in *some* medical evaluations. Are we certain where this paradox *always* takes us in physics? Also, we haven't even touched chaos theory directly.[24]

These are overwhelming arguments for believing Einstein's deterministic instincts were wrong. But one needs to look at other events in physics and mathematical tools for proofs to better understand Einstein.

Irreconcilable Theories: The Heisenberg/Schrodinger Story

Many irreconcilable differences in physics have been resolved to the envy of many broken marriages. Is it out of the realm of possibility that some sort of mathematical resolution could be found that shows both the Heisenberg/uncertainty and Einstein/exactness being right? A similar seemingly irreconcilable conflict regarding particle/wave was resolved mathematically during Einstein's life. Heisenberg was looking at atoms and accurately predicting a phenomenon as a particle. Erwin Schrodinger was calling the same phenomenon a wave and also achieving impressive results. What was the phenomenon - a particle or a wave? Sides were being taken throughout physics departments on whether Schrodinger or Heisenberg was right. Almost all physicists at the time believed that one or the other *had* to be wrong. The mathematician Paul Dirac resolved this paradox with his remarkable math creativity. He showed how these two apparently contradictory theories were, in fact, in harmony. Einstein was hoping his math talents might be able to resolve the apparently irreconcilable Heisenberg's uncertainty versus his own determinist view -- as Dirac had done with the particle-wave paradox. Is such a solution impossible?

Proof by Contradiction

There is even stronger mathematical rational for Einstein's stubbornness. The mathematical methods found in "Proof by Contradiction," and arguably, "Proof by Induction," can be examined to understand his optimism. Proof by contradiction is a powerful tool in theoretical mathematics, yet one can easily argue that Einstein would have missed Relativity if he had not boldly continued to explore our world beyond the boundaries of math's contradictions. Why should Einstein restrict his thoughts within the boundaries of "uncertainty theorems" when it was his wilderness exploring *beyond* the typical boundaries of

---

[24] Appendix E looks briefly at chaos.

"contradiction theorems" that took him to his most famous discoveries?[25] Which exploring is likely to be more fruitful at first glance: exploring beyond uncertainty or contradiction? His instincts might be proven wrong this time, but was he unreasonable in thinking he might overcome hurdles discussed by Heisenberg and Godel? No! Ridiculing Einstein's stubbornness shows a lack of appreciation for math's frailties and enigmas. Our daydreaming hero had an obsessive craving for mathematical wilderness exploring.

This is all fine and dandy philosophically, but was Einstein right in his deterministic views? Almost all who love to gamble will surely bet against him. This is why few books go deep into Einstein's later work beyond 1925.

Back to Modern Relativity and its Current Conventional Wisdom

Supporters of Einstein's deterministic views still exist in many physics and math departments around the USA. But a growing list of noteworthy physicists feel something big might be amiss. These talented physicists include Dan Sheehan, John G. Cramer, Keith Peacock, Ayman Abouraddy, Peter Woit, Jeff Kimble, Joao Magueijo and Lee Smolin, just to name a few. However, many other physicists, like Stephen Hawking and Brian Greene, have the more natural inclination that only modest tweaking of our current theories is needed to set things right. Today's relativity theorists tend to concentrate on complex multivariate "strings." Their thinking centers on trying to create new variables in hope of matching theory with observations. But, these new multi-dimensional variables made to fit experimental data and thought experiments are a mathematical double-edged sword. Strings with arbitrary dimensions and variables may describe a mystery solution – or alternatively – be nothing more than an accounting "quick fix" for an underlying false assumption.[26]

Moreover, modern relativity theorists love to explore many different shades of equivalence issues that, in some cases, might be never-ending circular logic of a typical Godel sentence, thinking they're closer to their goal since they've worked so hard, but in reality not that different from dogs spinning to catch their own tails. Ironically, relativity theorists tend to

---

[25] See appendix E for more on "proof by contradiction" and then review pages 65-66.

[26] Appendix A deals with a paradox – Relativity initially appears as a paradox. They both go in all sorts of directions. This Berkeley/cancer-option paradox appendix can give an appreciation for why the more complex multi-variate Relativity becomes volumes of equations – easily used to construct complicated mazes of broad orthodox certainty, based on an underlying foundation of a particular resolution for an apparent paradox.

cling tightly to c being the top speed possible, and the absoluteness of Cem. [27]

An obsessive conviction that no further revisions to Cem are possible appears especially ironic when the learning portal to Relativity is through Examples 1 & 2. Applying only Option 2 in Example 2 to derive $E=mc^2$ is suspiciously similar to the original incomplete thinking of Example 1.[28] Our original assumptions in Example 1 led us to change only time instead of time and space. How similar was our assumption that only time changed in Example 1 to only mass can be responsible in creating E/C^2 in Example 2? Assuming Example 2 is like Example 1 then the warping of space occurs. But is that a reasonable event for an intuitive sense of Cem? Sure, why not? But beyond accounting convenience is the unsure side of why. And if the warping of space occurs, why are we so certain that warping all stops there? What about that Option 1 we rejected in Example 2? How confident are we that we've accounted for all aspects of energy from all possible events?

---

[27] What Einstein and friends called "Spooky action at a distance," is a phenomenon that, at first glance, appears to travel much faster and differently than c. Light emitted from beta-barium borate can be split by a prism and then when one of the split beams is polarized, this causes an instantaneous polarization of the other beam – so much faster than the speed of light, as measured by our current abilities to calculate speed, that it 'appears' to be instantaneous...hmmm. Likewise, not detecting gravitational waves from pulsars, black holes and elsewhere might be explained by the "gravitational effect" traveling differently than as just c. Are some things interacting a lot faster than the speed of light? – A direct contradiction to Einstein's assumptions! The fact that both: some things appear to be traveling differently than light *and* reasonable challenges can be made that question the *absoluteness of Cem* is unbelievably bad news for intellectuals who have deified Einstein's assumptions as unquestionably true and complete. This is an analogous dilemma to the intellectual establishment of 1634 mandating ancient assumptions as correct and thus the earth is flat. Advocates challenging aspects of Relativity should realize the passionate allegiance at our learning institutions to their current teachings. One should not underestimate the subtle efforts to scuttle contrary thinking and how easy it is to be sidetracked by those who have spent considerable effort to learn the volume of complexity found within Relativity – not unlike the volume of statistical mazes of cranial profiling that Twain successfully overcame. To those who find this discussion troubling, fear not. C and our current Cem is the reality of many events around us. Einstein's assumptions are applicable insight; they just might not be as prophetically all-encompassing as most believe. Alternatively, to those who interpret this book as encouragement to try to experience Einstein's joy for trailblazing – have fun and good luck.

[28] Review Einstein's path to $E=mc^2$ on pages 66 and 71 if needed.

Exploring Beyond Conventional Wisdom and Relativity's Current Boundaries

The reliance on the absoluteness of conservation of energy and momentum (Cem) has created many vital equations in Relativity. Its usefulness is difficult to overstate, as seen in our two simple pool-ball examples, to Admiral Nelson's victories, to General Relativity's underlying foundation. But, Cem is dramatically different now from its definition in the 1800s. Are there no further revisions to be made? Are there no happenings beyond our current boundaries of Cem?

First, let's list why these questions are as vigorously ignored now as they were in the 1920s:

1) Classical & electro-magnetic physics always assume Cem. Why abandon this assumption for Relativity? It would be difficult to formulate many of traditional physics' most important equations without Cem. How many of our existing equations would be subject to review?

2) We assumed conservation of momentum when deriving $E=mc^2$ (however, itself very counter-intuitive to the events in our Example 2 – shades of a Godel sentence). One can logically argue that we also assumed conservation of energy. Therefore, Example 2 has a built-in paradox when not assuming Cem. Not only that, Relativity's Cem is clearly appropriate for countless observations.

Asking a physicist, "Is the assumption of absolute conservation of energy and momentum always true?" is similar to asking a 16th century ship captain in the middle of the Atlantic if he is sure his compass points exactly north all the time. One can appreciate why a seasoned physicist is alarmed by our doubts.

However, let's tweak our question by assuming our current Cem remains the norm under all but the most unusual situations. Can a deviation of our current definition of Cem be qualified, quantified and hence reincorporated into a revised Cem as was done by Einstein in 1916? This would resolve many legitimate concerns.

Where else might these speculative questions lead us? Unfortunately, Relativity is a poor vehicle to test the premise of further possible refinements for Cem. Relativity is inherently confusing. Simple math models in classical or electro-magnetic physics are better arenas to challenge our current understanding of Cem.

> "A novice might grasp what experts believe is unattainable."

> *- Anonymous*

Chapter 5

A Sphere: Dilemma with Even the Simplest of Objects

There are many unanswered questions in physics. One question often asked not that long ago was, "What makes the world go round?" Money or love? Which song is right? Neither answers the question to a physicist's satisfaction.

Also, why do outer electrons jump from one energy level to another specific level in an atom without being able to occupy any of the intermediate energy levels? Why are electrons and protons spinning like tops? Why are electron sizes so similar and exact instead of being many different sizes like planets and stars? These facts exist and are taken for granted...but why is it the way it is?

On the opposite size of physics, what existed long before and right before the big bang? Why would something so dense go bang? Explaining the big bang is very difficult without some very creative Cem accounting techniques dwarfing the accounting creativity of businessmen enjoying vacations in the drab surroundings of federal prisons.

This speculative chapter is a springboard to answering some of these questions. It's taking a gamble and placing a bet against conventional wisdom.

Arguably one of the simplest shapes of mass for evaluating gravity is a sphere made of fluid homogeneous particles of uniform density. Can a simple math proof describe this simple object as a possible dynamo? What are its implications?

The sphere we'll examine is made up of a fluid similar to water. But this sphere has unusual characteristics. First, it is very simple mathematically so any paradox is not cluttered. Second, the closer the particle is to the sphere's center, the more it wants to float to the top because the particle above it is heavier and wants to displace it. We will quantify an unstable equilibrium that loses stability once impacted by an outside force.

Is this dilemma possible? If so, what does this mean for inertia and the absoluteness of the conservation of energy and momentum? Does this mean that some types of mass accumulation are impossible just like an infinite number of possible movements of pool balls after a collision are also impossible? If so, what are some of the many other impossible accumulations and why are they impossible? How big of a role does this play in the discreetness around us? Or is this the simple start of something even more complicated than just restricting accumulations?

This chapter will give a feel for a type of math proof that needs only a pen, paper and imagination. A moderate level of theoretical math experience is needed to follow the proof but the prose is for general reading.

This proof is presented in hand-written form to emphasize its scientific wilderness and the ability to tackle conventional wisdom without computers. It assumes point mass and the widely accepted gaussian concept that exterior shells' net gravitational force is zero (we are not excluding pressure).

Force toward sphere if m is on surface of M where M is mass of sphere.

$$F = \frac{m(GM)}{R^2}$$

R = radius of sphere

G = gravitational constant

Volume of sphere = $\frac{4}{3} \pi R^3$

Assume ρ = density is constant

$\therefore$ Mass of sphere is $\left(\frac{4}{3} \pi R^3\right) \rho$

$$F = \frac{GMm}{R^2} = \frac{Gm \frac{4}{3} \pi R^3 \rho}{R^2} = Gm \frac{4}{3} \pi R \rho$$

Assume sphere is of uniform density with particles of same mass equidistant from each other.

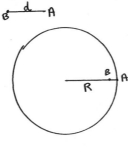

If A is a particle on the surface of the sphere and B is the closest particle in the direction of the center of the sphere then ΔF_{AB} is the difference in gravitational attraction toward the center of the sphere.

$$\Delta F_{AB} = G \frac{4}{3} \pi \rho m R - \left(G \frac{4}{3} \pi \rho m (R-d) \right)$$

$\Delta F_{AB} = G \frac{4}{3} \pi \rho \, dm$

Neglecting friction and chemical bonding there is a discreet measurable force whereby the outer particle will want to change places with a particle on the next inner shell.

If we sum up the ΔF between all such particles, $A \& B$ of distance d apart along a vector originating at the surface of the sphere and extending toward the center we have:

$$\Sigma F = \sum_{i=1}^{n} G \frac{4}{3} \pi \, d_i \, \rho m_i \;\; =$$

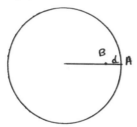

$$= G \frac{4}{3} \pi \rho (d_i m_i + d_2 m_2 + \ldots + d_n m_n)$$

but $m_i = m_1 = m_2 = \ldots = m_n$

and $m_i = \rho d_i^3$

$$= G \frac{4}{3} \pi \rho (d_1^4 \rho + d_2^4 \rho + \ldots + d_n^4 \rho)$$
$$= G \frac{4}{3} \pi \rho^2 (d_1^4 + d_2^4 + \ldots + d_n^4)$$

but since $d_1 = d_2 = d_n$

$$= G \frac{4}{3} \pi \rho^2 n d^4$$

Now let's say $y = d = u = 1$ for unit of measurement.

$$= G \, {}^{4}/_{3} \, \pi \, \rho^2 \, n \frac{1}{y^4}$$

$n \cdot d = R$ and $d = \frac{1}{y}$

so $n = \frac{R}{d} = Ry$

$$\boxed{= \pi G \, {}^{4}/_{3} \, \rho^2 \frac{1}{y^3} R}$$

This force applied over a surface area (lbs./sq. ft.)

$$= \pi \, G {}^{4}/_{3} \, \rho^2 \frac{1}{y^3} R \cdot y^2 \implies F\!/_{\substack{sq. \\ unit}} = \pi G \, {}^{4}/_{3} \, \rho^2 \frac{1}{y} R$$

Now look at when the accumulative flipping equals gravitational attraction of mass m at a distance R from surface and R' from the center of the sphere.

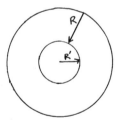

$m \, G R' \, {}^{4}/_{3} \, \pi \, \rho = F_G$ sphere

$G R \pi \, {}^{4}/_{3} \, \rho^2 \frac{1}{y^3} = F_{Flipping}$

For what R, R' are these forces equal?   Set $F_{G \, sphere} = F_{Flipping}$

$$m \, G R' \, {}^{4}/_{3} \, \pi \, \rho = G R \, {}^{4}/_{3} \, \rho^2 \frac{1}{y^3}$$

cancel $G \, {}^{4}/_{3} \, \pi \, \rho$

$$R' m = \rho \frac{1}{y^3} R$$ but $\rho = \frac{m}{d^3} = \frac{1}{1/y^3} = m y^3$

$$R'm = my^3 \frac{1}{y^3} R \qquad R'm = mR$$

$$\Rightarrow \quad R' = R$$

At the midway point of a sphere the flipping of a particle will equal its gravitational force.

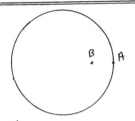

F_G = gravitational force
F_c = centrifugal force

Look at equator with sphere spinning

$(F_{GA} - F_{CA})$ = net force on A

$(F_{GA} - F_{CA}) - (F_{GB} - F_{CB}) = \Delta F_{AB} - (F_{CA} - F_{CB})$
but $F_{CA} - F_{CB}^{*} > 0$ hence $\Delta F_{AB} - (F_{CA} - F_{CB})$
on equator is less than at poles at equal distances from center. For example:

Pressure equilibrium on pole to equator would be:

$$B_i = \sum_{i=1}^{n} G^{4/3} \rho \pi R_i \, m_i = \sum_{i=1}^{n''} (G^{4/3} \rho \pi - \frac{4\pi^2}{\lambda^2}) R_i'' m_i$$

where $nd = R$ @ pole, and $n''d = R''$ at equator. λ is the rotational variable for centrifugal force.

This would be the start of many variables converging to a ΔE min from our current definition of CEM. Our specific example would continue with a further summation where $B_{sum} = \sum\limits_{i=1}^{n} B_i$ as B_i moves along the rotational axis to the poles.

Let $B_{sum} = B_1'$

Is a spin associated with this flipping instability set to a minimum?

i.e. Is ΔE min $\subset (B_1' + B_2' + \cdots + B_n')$ where $B's$ represent all other factors at play?

Is this pure fiction or relevant to our universe? What are our options?

Option 1) The proof suggests the possibility of a dynamo. But despite its compelling simplicity, it may contain an unknown flaw, as does "what goes up, must come down." However, the same can be said about the assumptions for General Relativity.

Option 2) Alternatively, nature may just mandate that such a sphere cannot exist because it would violate the absoluteness of Cem. If the simplest of objects must be avoided to maintain absoluteness of Cem, then there are likely many forms of physical accumulation, beyond our known kinetic restrictions, that are also banned from reality because of the absoluteness of Cem. This might explain much of the curious discreetness of QM. Or, if such forms of physical accumulation are not banned, are dramatic revisions of Cem needed wherein our current conception of Cem is merely the most desired state within the reality of more complicated events? Dissecting this possibility in Options 3 & 4 offers even more exciting possibilities.

Option 3) The movement or turbulence (if it does exist) in this sphere may be offset by loss of energy elsewhere. For example, a decrease in any orbital trajectory energy with other bodies may "balance out" this "new" energy. Also disturbing is no work is actually performed since one particle merely changes place with another – so is any "balancing out" from elsewhere needed? So we have stationary particles becoming kinetically dynamic while performing no net work, at least in the manner work is currently defined as a reduction of potential energy. Also, if our current Cem holds under these dynamos, then inertia needs to be redefined to accommodate these events...something has to give.

Option 4) The sphere is a dynamo where change of motion is not totally negated by the removal of energy/mass elsewhere. If dynamos as described in our sphere do exist, with no loss of energy elsewhere, then a revision of Cem is needed. A counteracting *force* to stop such a dynamo can be estimated – but it is slippery to incorporate a timeless force into Cem.

Options 2, 3 and 4 suggest possible reasons for some of the discreetness in quantum mechanics. If either Option 3 or 4 is reality, then an argument for practical perpetual motion machines can be made even beyond the boundaries of "Maxwell's demon" as discussed in the next chapter. What efforts would nature take to deliver one atom to our current thinking of Cem versus allowing a more complicated concept of Cem – a complex minimal diversion to our current Cem that would still mandate discreetness – like that found in QM? Rotation, size limitation, and trajectory confinement are three of nature's slickest ways to accommodate our current thinking of Cem minimizing this dilemma. Hmmmm.

In summary, let's revisit the infancy of Cem. Was Newton's genius in stating, "Every action has an equal and opposite reaction," more an insight of nature, or a useful description, noting that essentially any event can be defined as a whole and hence divided into equal halves for math convenience? This eloquently simple observation enabled remarkable scientific discoveries. As the focus in physics turned to energy, the parallel and more powerful thinking of conservation of energy evolved. The conservation of energy and momentum has been redefined before. We have seen how its usefulness is difficult to overstate. But there is a subtle difference between Newton's statement and the concept of absolute conservation of energy. Making two halves out of a whole is simply a useful method to describe an event. Saying something is always there never to increase or decrease is an assumption based on observation and faith.

There have been many new observations since 1916. Can our current definition of Cem honestly describe the events leading to the Big Bang? Are there other events in our universe that also cannot be described within constraints of our current Cem? Why should reality necessarily be so limited? An ability to accurately measure stable time was needed to develop classical kinetics. This stable time measurement is now often replaced by the variable time of Relativity. In much the same way, one type of Cem was necessary for classical physics and thermodynamics while a more expanded Cem was needed for Relativity. Should we look beyond our current strict view of Cem in an effort to expand our knowledge of reality? Have physicists glorified the essential Cem to **unrealistic** heights, and turned into one of the most respectable cults the world has ever seen?

We have too many unanswered questions and are contorting too many other variables to demand that Cem should not be explored as also being more flexible.

Finally, what happens to the vanishing gravitational force that appears to go unused in physics? This question applies for both the classical gravitational shells[29] and gaussian boxes in electro-magnetics that so fascinated Tesla and Twain. Do these annulling forces slowly accumulate until there comes the proverbial "straw that breaks the camel's back"? Or can they cancel out each other forever? Either answer adds a complex enigma to the concept of potential energy and hence, also for conservation of energy. Is this also where one should explore if wanting to discover a big piece of the puzzle as to why the big bang occurred?

Are annulling forces that so fascinated Twain a start toward answering today's awkwardness in Einstein's physics?

---

[29] See *Physics I*, Halliday and Resnick, second edition, page 399, for the mathematical explanation.

Chapter 6

Smart or Foolish: Trying for the Unattainable?

Perpetual Motion Machines are Impossible!

Are perpetual machines impossible? It all depends on how you define one. The 1775 meeting of the French Academy of Sciences that proclaimed their impossibility had a very different concept of perpetual motion machines than how they are defined by physicists today. The concept of early perpetual motion machines was a device you could put in a carriage and take to anywhere in the world, take it out and voila, it would start a rotating device that could be harnessed to a fan or flour mil. This device would save man and beast alike from the chore of doing work. The French Academy of Science was growing tired of harebrain ideas and cheaters presenting devices that on close inspection didn't work. Besides, there was very good logic suggesting that such a device is impossible.

This proclamation came on the heels of a con man extraordinaire whose given name was Johann Bessler. He had created a perpetual motion machine as described above and had many famous & knowledgeable men intrigued. Bessler teasingly showed the machine working but was forever postponing an unveiling of its inner workings to the public. His excuse for postponements were varied but one of the more noteworthy had to be his lack of spare time while changing his name to Orffyreus – as in rhythms with Orpheus, the god who crafts mesmerizing songs. Sir Isaac Newton's comment, "The seekers after perpetual motion are trying to get something from nothing," was too late for eager optimists unfortunate enough to fall under Orffyreus' spell by witnessing his machine that seemed to work indefinitely. Few who invested into Orffyreus' miracle ever admitted that they had been conned – a testament to human nature. How one would love to go back in time and view the scientific seductions of Orffyreus, the Don Juan of perpetual motion.

Johann Bessler may have failed at actually creating perpetual motion, but at least his con-artist ways as Orffyreus set the French Academy of Science into motion. Their long ago proclamation now appears to be perpetual. The Academy's 1775 exact words proclaiming that perpetual motion machines are impossible were:

"The construction of a perpetual motion is absolutely impossible; even if the effect of a moving force is not destroyed by friction and resistance of the environment, this force <u>cannot produce more reaction than one equal to its action</u>. If the effect of a finite force continues, it is necessary that this effect is infinitely small after a finite time. If friction and resistance are ignored by abstraction (which in nature is not the case), an object that got a movement once, would remain in perpetual movement, but not act on other objects, and thus the perpetual motion would be completely useless in regard of the claims of the inventors. This part of research has been inconveniently expensive, it has ruined more than one family & mechanicians, who otherwise would have been of great use, have wasted their luck, their time and their genius."

The modern day explanation of why perpetual motion machines are impossible is much more specific and technical. Today's education of physics uses the colorful history of eccentric perpetual motion to display the beauty of Cem and thermodynamics. We'll use this same colorful history of quack physics to paint a picture of optimistic uncertainty.

An examination of one machine will show how scientists living in 1770 versus today would point to different flaws disqualifying it as a coveted perpetual motion machine.

A Tide Machine

Imagine a large container, like a huge cup, that can hold a large amount of water. Now assume it has two openings one fairly large and the other comparatively small both near the bottom. Now place simple floppy check valves on both openings where one check valve lets water into the container through the larger hole and the other lets water out the smaller opening. Place a small electrical generator turbine on the small hole and send the current to a battery for storage. Now place this container on an ocean shoreline right in the middle of the tidal zone. So as the tide comes in, the container is filled. As the tide goes out the container is emptied and as the water leaves through the small hole a turbine generator sends electrical energy to the battery to be stored until used by a fan or other device. The battery is being charged day in, day out for the foreseeable future. This occurs twice a day as the tides go in and out.

Better yet, put turbines on both openings so as the water goes in it also generates electricity that can be stored in another battery.

Finally, a little tinkering with the valves and this tide-tank really kicks into high perpetual motion gear. One creates a useful dynamo if clever tripping mechanisms are created to keep the valves closed until the tides reach their high/low points. This will increase the practical electricity generated many fold.

Question: Why isn't this a true perpetual motion machine?

Answer, 1770 version: This device works only near the coast. We envision a machine that can work wherever man resides.

Answer, today's version: This machine will not work indefinitely. Entropy described in the Second Law of Thermodynamics mandates the stopping of this machine. There are other more philosophically involved arguments why Cem mandates that this machine is not perpetual motion, but entropy is usually the first argument cited as why it does not cut the mustard as a truly perpetual motion machine.

What is entropy? Entropy is the word coined to sum up the Second Law of Thermodynamics. This law states that hot things, like a cup of coffee or a red-hot block of steel, slowly cool down to their surrounding. Pretty basic. Outer space is very cold and this cooling of our hot cup of coffee is extended to the universe. This trend has an endless number of broad observations confirming its obvious existence of hot objects cooling to the level of their immediate surroundings. Equations can be crafted using flavors of entropy that help design efficient power plants. Efficiently utilizing **entropy differential** of heat is **the key** to a well-designed power plant – just as the tide machine having the valves only being opened during high/low tides *is the key* to getting a lot of useful electricity out of our tide machine.

This basic theory of entropy can be thought of as a road that is always going downhill to colder. There might be small bumps in the road or side roads that don't appear to go down, but over the long run everything merges back to this imaginary road that just keeps going to colder. A real example is our solar system with our hot sun at its center. Our sun will eventually burn out. $E=mc^2$ gives us reasonable estimates on when this will occur. Entropy mandates that Earth will become a very cold wasteland once the sun is no longer supplying solar heat...so basically, everything on earth will eventually run out of gas.

Thermodynamics does not offer any long-term counter balance to this factual phenomenon and, in fact, proclaims that no counterbalance exists! Entropy will eventually over-ride all. Believers that entropy is making our universe forevermore and beyond colder are members of a secular form of religion – kinda the opposite of those believing in an eternally *hot* hell.

This tide machine is a very small speck in the road of entropy for those convinced that all perpetual motion machines are impossible. But is this more a philosophical excuse for technical denial that perpetual motion machines might be possible? After all, none of us are going to last forever so why must a perpetual motion machine? Is our conventional wisdom about perpetual motion oblivious to reality, and more a face-saving mantra for the long gone souls of 1775?

We will shortly present a few interesting ideas of possible tide machine semi-equivalents that, if workable, could be used all over the world. But to get a clearer picture of our direction of thought, we will first describe a few more believable variations on the concept of perpetual motion. It is easy to understand how these mundane machines circumvent the entropy barrier to perpetual motion. They generate heat warmer than their environs and can perform work indefinitely with respect to human existence. Are these <u>practical</u> perpetual motion machines – minus the mumbo jumbo of overly restrictive thermodynamics arguments claiming machines constantly doing work are impossible?

The Polar Bear Dilemma to Entropy Arguments Against Perpetual Motion

Generations of polar bears are an interesting anomaly in the road of gradual disbursement of heat (entropy). They live in an environment almost always at a lower temperature than their body heat. Polar bears accumulate energy stored in cold-blooded fish having lower temperature than their own bodies. They have maintained this anti-entropy existence and will likely continue to counteract entropy for the foreseeable future with respect to man's existence. Their thickheaded brains have successfully circumvented the entropy barrier to perpetual motion for thousands of years.

A self-feeding bulldozer running indefinitely by burning the wood it accumulates in a rainforest is a believable mechanical device similar to an artic polar bear. Most engineers believe such a device could be built with today's sensors and robotics. This device and animals clearly work within the realm of Cem and thermodynamics since they are using energy stored from the sun that happen to be temporarily cold. Long-term entropy is not broken, just temporarily diverted.

So, we have tide machines, polar-bear-like machines chomping away at the rainforests to the chagrin of environmentalists ... if we wanted. How many variations of these sorts of devices are possible, and are some of the hare-brained impossible machines possible? In other words, do the boundaries of impossible perpetual motion machines need re-evaluation?

Mobil scavenging machines that work indefinitely are a given. The tide machine is stationary. It utilizes gravitational variation and the oceans' dual action of accumulation and delivery of energy from this variation. Maxwell presented a stationary "Maxwell demon machine" that challenged the concept that perpetual motion machines cannot be made. His machine selectively allows only faster moving particles into his device. Hence his device contains higher kinetic energy than its surroundings and therefore can be made to do work. But the 1775 edict largely inspired to prevent Orffyreus-like con jobs is given homage by almost all physicists while the possible insight of the remarkable Maxwell is rejected outright.

Most proposals of modern Maxwell-like devices are stationary and use the concept of accumulating surrounding energy and then utilizing this accumulated energy in an efficient manner to create a dynamo. They rely on the loss of heat (or energy) in the immediate surroundings being replenished by the heat in the greater surroundings. Ironically, entropy sustains such a device rather than mandates its impossibility since it's the surrounding heat enveloping it that prevents its eventual halting from lack of energy. The *overall* direction of entropy will not be changed nor is there a conflict with Cem, if, in fact, such a machine can be made.

But now we have a speculative sphere from the last chapter that suggests constant dynamos **might** be a fact of life all around us. This directly contradicts the widely believed claim that such dynamos can't exist. Entropy's absoluteness that all will halt is also contradicted by this sphere. If this sphere is part of reality, then practical perpetual motion machine proponents have even more reason to be optimistic. A little over a century ago, starry-eyed optimists argued, "Birds can fly so why can't we make flying machines do the same?" Their modern counterparts might soon argue, "Natural dynamos exist, so why are we so sure we can't utilize or make them?"

Entropy's Subtle Flavors Have Broad Theoretical Value for Conventional Wisdom

It's remarkable how this simple observation of hot things cooling to their immediate surroundings has expanded into so many interesting, useful, yet restrictive corollaries. One of entropy's corollaries to hot spreading to cold is that the "disorder" in the universe is constantly heading to more "order." This is a popular topic for many physicists. Theoretical discussions of entropy often lack exacting algebra equations - but not vagueness - like beautiful music that can be heard, but not grasped. This makes nuances of entropy great umbrellas for protecting conventional wisdom from contrary experiments and theory – or for proposing new concepts.[30]

But a reality check is needed to this discussion attacking the power of entropy. No successful stationary working perpetual motion machine of Maxwell's demon persuasion has yet been made despite many attempts.

To re-emphasize our last warning for optimists before moving on, excerpts of the wise last point made by the 1775 Academy are in order:

> "This part of research [those who tried to make
> perpetual motion machines] have wasted their luck,
> their time and their genius."

Modern translation: Keep your day job, your money in the bank and don't become too obsessive with a perpetual motion machine dream!

The Van der Waal Opportunity

There are micro-vibrations created by Van der Waal forces in the miniature world of quantum mechanics. These eternal vibrations in the molecular world exist throughout our planet. They are well documented. The existence of Van der Waals' vibrations is as certain as the tides in our oceans. Miniature machines working on the molecular level are closer to reality than most could have even imagine a couple of decades ago. Could our ability to create smaller and smaller miniature devices create an opportunity to use Van der Waal perpetual vibrations as the driving force in mini current creating devices? The tides vibrate with a frequency of 12 hours and can be easily manipulated to deliver energy for use to mankind. Can the same be done with the fast vibrations driven by Van der Waal forces that exist in molecules wherever humans reside? Some hope that charged endpoints of a very small semiconductor chip will vibrate because of surrounding Van der Wall forces much like strings in a guitar will vibrate from noise with the right frequency. This vibrating chip can then be used to create a current, like rotating (or vibrating) charges of our everyday generators are used create electrical currents for household electrical use. Unfortunately, devil's advocates can make fairly convincing arguments that the randomness of the vibrations neutralizes any possibility of creating a useful current. But are such arguments any more

---

[30] Our second very simple billiard example in Chapter 2 demonstrates why Cem is so important and why the concept of an expanding momentum is rejected outright. Yet this 'wrong' concept of momentum expanding is like entropy...hmm. Is the wrong answer for predicting billiards movements that maxes out to $(n^{1/2}-1)$ a link to entropy? Is this insight for today's incredibly useful Cem conflicting with the themes of Chapters 4 and 5?

conclusive than arguments presented for the truth of "What goes up, *must* come down"?

Papers in Appendixes F and G look at different types of heat sink and miniature charged endpoints as practical perpetual motion machine concepts. Computer simulations in one paper, and the construction of a device in another paper, claim to demonstrate very promising results.

These generators, as presented in their detail, may be very tempting. But do they really work? *Only a physically working machine that can also be made by others will convince pragmatic doubters*. Also, it would be remiss to leave without mentioning that other dynamo concepts sounded so logical before they were built, but upon construction they just didn't work. A few of our examples will show reason for caution.

The "Bobbing Bird" Uses Entropy to Its Advantage

The "Bobbing for Water Bird" is an amusing toy analogous to small twirling steam toys of a millennium ago. It appears to effortlessly "drink" continuously – under the right circumstances. It does this by using internal evaporation and condensation of a volatile at room temperature, external wicking/cooling requiring the room containing the bird to have a low relative humidity for evaporation. The construction includes a clever tube and reservoirs deceptively difficult to design to ensure operation. Just the right leveraging of the bird's neck is needed. Its operating temperature range is too small for worldwide use and the water being "sipped" has to be at just the right level. But, if all the above  exists then...voila! Looks like a <u>practical</u> perpetual motion machine that is practically practical!

The entropy action (variation of temperature which creates a drier atmosphere) of our planet is largely responsible for making this bobbing bird possible. Hence, neither entropy nor Cem are in conflict with this device. Which brings us back to our philosophical debate: What is or is not a perpetual motion machine? At what point are we playing a game of philosophical physics semantics cloaked as intellectual analytical discussions?

Does this bobbing bird satisfy the simple 1775 definition of perpetual motion machines? You decide for yourself.

Whatever your answer, the bobbing bird is not necessarily good news for optimistic perpetual energy machine inventors; it is of no economic value except at novelty and toy stores. But, then, so were twirling steam toys in olden times.

Even if possible, could a clever Van der Waal machine compete economically with our existing sources of useable energy?

Other Entropy Concepts and Reasons for Caution

Loschmidt Logic

Heavyweights like Poincare, Mach and many others have questioned the degree of importance of our laws of thermodynamics as they apply to our universe. But we will look at Joseph Loschmidt's challenges to thermodynamic laws as his concept fits our theme of semi-resolving uncertainty. The great Loschmidt – Boltzmann debate a century ago pitted entropy against possible perpetual motion. Loschmidt is a famous scientist in Germany but his famous discovery was attributed to another in the US and Britain partially because his birthplace was on the losing side of World War I. The debate turned on whether a greater barometer pressure means a continual increase of temperature. Loschmidt said yes and Ludwig Boltzmann, Maxwell and many others said no. The Boltzmann argument was that entropy prevailed. His thinking carried even more weight within the science community because, if Loschmidt was right, certain types of "impossible" perpetual machines would be possible.

The Loschmidt-Boltzmann debate is a great example of how scientific debates should be conducted. Egos were never bruised to the point where friendships were broken. Boltzmann was the main speaker at Loschmidt's memorial service. Both sides had interesting, reasonable logic to support their position. Yet everyone believed the contradicting conflict meant one side would overpower the other.

Loschmidt's thinking was that the high school thermodynamic chemistry formula;

$$P_1V_1/T_1 = P_2V_2/T_2$$

is a major factor in why temperatures tended to be cooler at higher elevations and remain so even despite the fact that entropy should be dissipating this difference if the air was stagnate. Like our theoretical sphere dynamo of the last chapter, there are mountains of evidence that can be selectively chosen to support this theory. Likewise Loschmidt's opponents need only to head to a very tall building with a gas enclosed in a long vertical cylinder and note a greater pressure at the bottom of the tube does *not* also mean a greater temperature. This would show that entropy prevails and is a powerful entity in the universe. Boltzmann could have also argued entropy creating a uniform temperature means still another method of creating perpetual motion machines was impossible. Today's fans of Loschmidt's thinking are a much smaller minority than Einstein's determinist followers. Boltzmann's logic is self-evident from observations: heat dissipates to cooler surroundings. Loschmidt's logic is easy to show.

We will start with $P_1V_1/T_1 = P_2V_2/T_2$. This equation is simplified to $P_1/T_1 = P_2/T_2$ since the volume is the same in Loschmidt's thinking. So,

$$P_1/T_1 = P_2/T_2$$
this implies $\rightarrow$ if $P_2 > P_1$ then $T_2 > T_1$.

Therefore, air at lower elevations has a higher temperature. But do these equations apply to this type of stagnate air conditions? Over time, does entropy make this equation irrelevant in the atmosphere as it does in a compressed cylinder? (The answer favoring Boltzmann is more intuitively right today than it was 100 years ago.) Loschmidt strongly believed that a larger pilot light remained for those gasses having higher pressure. Twain's close friend, Nicolas Tesla, conceived power-plants in the early 1930's which were actually built decades later using temperature differentials of oceans' depths as a source to generate electricity. These projects correlate well to perpetual motion machines being possible if Loschmidt was right. Appendix H, having a strong presentation (biased) for Loschmidt, will look at the devil in the details argued by Boltzmann and Loschmidt 100 years ago. It will discuss how molecular interaction results give reason for Loschmidt's thinking. Are we sure Loschmidt was completely wrong?

Before moving to a different topic – a bee in a bonnet, assuming entropy prevails – we must ask the question: Is Loschmit's temperature/pressure differential responsible for more turbulence on our planets and in the moons of Jupiter than current conventional wisdom asserts?

The Stevin Inclined, Sponge-activated Machine

The epitaph of Stevin over 400 years ago is one of the earliest recorded documents of intrigue over conservation of energy and work (see drawing). He was obviously enchanted by the fact that the weight of few balls could counter weigh many balls and prevent perpetual motion.

A clever more modern inventor proposed tweaking the triangle slightly and using this gravity equilibrium to his perpetual motion machine's advantage. He combined two gravity balanced rings to make a hybrid gravity unbalanced ring, which would cause the ring to slowly turn.

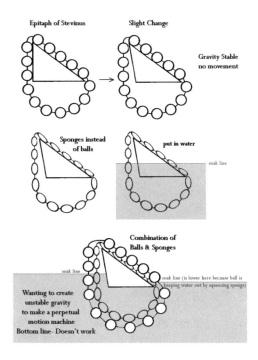

The concept was putting the bottom of the triangle in water and placing another ring this time made of sponges within the ring of balls. This way the balls on the incline would partially squeeze out water that had wicked up into the sponges, thereby creating what was once a gravity balance to a gravity imbalance (see drawing illustrating perpetual motion logic). The chain of ball and sponges would start turning if friction could be kept to a minimum – not bad logic.

As machines were being constructed, counter advocates were already arguing that the machine is not a true perpetual motion machine because the absorption ability of the sponges would be worn out over time and is only running on the energy stored within the sponges to absorb water.

This wearing (i.e., work) keeps Cem valid. Also, entropy and the sponge losing absorbency means the machine would have a very finite life.

The counter-advocates ended up not needing these philosophical arguments because, to our knowledge, no sponge-like machine has been made to rotate. This sponge machine shows the degree to which convincing arguments can be made for rotating devices that just don't work when constructed.

One argument often absent from both pro & con gravity imbalanced perpetual devices is the fact that $F=GM_em/R^2$ means equal mass objects will not have equal force if not an equal distance to earth's center. The mass closest to the center has a very small yet discretely greater force than those of higher elevations, and hence, is braking any rotation to upward elevation on the earth's surface (an opposite direction opposing force to that in Chapter 5 which is *inside* an "earth" with *uniform* density). This fact helps neither opposing argument regarding Cem, because it 1) clearly works against perpetual motion machines and 2) distracts from showing the beauty and power that conservation of energy has over physics.

Feynman and Teller

Lecture notes from the talented theoretical and teaching physicist Richard Feynman start with a basic physics introduction. Then he immediately pounces on conservation of energy - the impossibility of gravity-induced perpetual motion machines - with zeal. It's his first hot topic for bright-eyed freshmen entering Cal Tech. Why? Possibly he was tired of critiquing bad perpetual motion concepts from optimistic freshmen. He hoped to extinguish contagious false ideas before they're started. That way students will remain focused on more promising concepts in physics.

What is interesting in Feynman's discussions is he chooses to omit the braking force of delta gravity for identical masses having different elevations as squashing gravity driven perpetual motion concepts. He instead chooses a conservation of energy loosely tied to a reversible/irreversible entropy/thermodynamic-like argument. He goes on about pros and cons of gravity perpetual motion machines, "it is a beautiful line of reasoning...[but the impossibility of gravity perpetual motion machine] It turns out experimentally, in fact, to be true."

But does he abandon hope of an endless supply of useful energy for mankind? Hardly!

He ends his conservation of energy chapter, "With 150 gallons of running water a minute, you have enough fuel to supply all the energy which is used in the United States today! Therefore, it is up to the physicist to figure out how to liberate us from the need for having energy. It can be done." This is encouragement, not ridicule, for optimists eager to invent a <u>practical</u> perpetual energy machine.

Feynman was talking about controlling thermonuclear fusion. Edward Teller was a big proponent of two uses of thermonuclear fusion. The first was the hydrogen bomb to vaporize one's foes by the millions and the second was as an electrical power source so great that everyone in the world would have more than enough power – relieving a major economic tension for using Teller's first fusion invention. The first came to be in 1952 and the second is still mired in a technical/political morass. Hmm...too bad it wasn't the other way around. These power plant concepts usually use the heavy type of hydrogen isotope called deuterium that exists by the millions in every glass of tap water. Deuterium when fused creates helium and energy in the form of neutrons and/or gamma rays. Feynman's 150 gallons per minute of water to power the USA gives a sense of the energy released by fusing these hydrogen atoms.

This perpetual motion tale reminds us of "Goldilocks and the Three Bears." We have only extremes of bobbing birds that can't possibly create enough energy for anything useful and fusion that gives us way too much energy for mankind's current abilities to control. Like porridge *too hot* and porridge *too cold* – but so far we have no practical perpetual motion machine that's *just right*.

Summary of Entropy & Perpetual Motion Machines

An example of resistance to new thought in physics today is the teaching of entropy and its possible boundaries of applications. How long after 1916 has entropy been presented in many thermodynamic classes as an "all-encompassing law"? Does entropy eventually overcome all? Let's assume, $E=mc^2 \rightarrow E/(c^2) = m$, which means that mass is a form of energy. Masses are attracted to each other by gravity. Wouldn't the fact that mass is a form of energy and accumulates create possible broad caveats far beyond polar bears for the all encompassing applicability of entropy in astrophysics and elsewhere? The big bang and black holes have been considered a given in most physics departments for decades.

Since the big bang happened once – why is it assumed that'll never happen again?

Entropy obviously exists...but, are there much more important counterplays than polar bears' existence?[31] What is entropy's role for cosmology?

---

[31] Proof by Einstein's second least favorite type of theorems: proof by a contradiction. The Second Law = heat over the long run *cannot* accumulate.

Are black holes and at least one big bang part of a yang to the yin of entropy?

But now let's go full circle and use entropy's obvious existence to bolster our theme that questions conventional thinking of Cem. Both the math of an expanding momentum and a conserved momentum share the same space on pool tables. The conserved momentum has always been the right choice for predicting exactness and advancing our knowledge of physics...until we note that the expanding option is remarkably parallel to entropy. [32] Is this a useful perspective for more complicated Cems?

Finally, the conventional wisdom of 1775 regarding perpetual motion is easily countered by the more obvious. Two examples:

Superconductivity was a surprise and a marvel when first discovered. No resistance or friction is detected at all. On today's earth, work must be done to keep the temperature low enough to continue this perpetual motion. But what would future intelligent life on our planet think when earth is much further down the road of entropy and has a mean temperature is 5.4 degrees Kelvin? Superconductivity would make this old-fashioned impossibility of things rotating indefinitely look more like the natural state of events!

Today's solar panel coupled with a battery, fan and taken back to the French Academy of 1775 would cause them pause to declare a forever rotating device impossible. The cool breeze from the fan would give the science delegates comfort from their sweat as they pondered a different philosophical question. Self-propelled machines have been constructed that remain operating for decades and the boundaries of similar devices have not been fully explored – the counter is simply antiquated capricious conventional wisdom.

The entropy barrier to practical perpetual motion machines can clearly be circumvented.

Proof: There are many forms of energy; from mass (i.e. $m = E/(c^2)$), to potential energy, to kinetic, to heat; hence mass being a potential form of energy is a potential form of heat; mass under certain conditions accumulates; therefore, there exists a form of potential heat (i.e. mass) that accumulates; => <= QED. (Obviously speculative proof until accumulation of hydrogen/helium can be *proven* to create new stars, and/or massive black holes can be *proven* to go bang.)

[32] Review pages 57-59 for an example of the different momentums and footnote summary on page 94.

General Summary

A Discoverer's Guide:

Free time daydreaming and wilderness exploring has its advantages.

If you are looking for action – don't forget to check out the boundaries.

Education:

Cherish a child's creativity and beware of educators who think it's nothing to extinguish it. Medicating our over-active daydreamers to maintain a disciplined calm is today's hidden crime in our classrooms.

Teaching is a never-perfect balancing act of conveying conventional wisdom while trying not to suffocate a student's gift for creativity. Over-emphasizing one brings peril to the other.

We all stand on the shoulders of others. But are we standing on and looking over accurate conventional wisdom when trying to peer beyond the known? How often do our mentors guide us in the wrong direction – like the teaching of the world as flat for over a millennium and the zombie teachings of Karl Marx that mesmerized so many academic intellectuals for decades?

Twain and Einstein got directly to the point:

"I have never let my schooling interfere with my education."
- *Mark Twain*

"The only thing that interferes with my learning is my education."
- Albert *Einstein*

Statistics:

Statistics have their limitations...unless you're a pathological liar.

Physics:

The simple sphere in Chapter 5 creates a quandary where entropy is but a small fraction of the intrigue.

Maxwell, Da Vinci and Bernoulli all believed that perpetual motion machines were possible. Modern physicists claiming Feynman did nothing but pooh-pooh *practical* perpetual motion machines need to reread his book. Questioning the certainty of a particular definition of Cem has not been done by only fools. The last dramatic revision of Cem was Einstein's almost 100 years ago. A laundry list of unexplained events such as: the spinning earth, electrons and protons; turbulent planets of Jupiter, Saturn and their moons; many aspects of quantum mechanics discreetness; superconductivity; and finally, the big bang – justifies questioning whether our current conventional wisdom is all encompassing. Clinging too tightly to Cem for its inherent accounting convenience might be impeding science.

While the overwhelming majority of today's physicists consider our doubts as obviously wrong, there is a glimmer of hope.

"The rarest and most valuable of all intellectual traits is the capacity to doubt the obvious." - *Albert Einstein*

Appendix A

Quantifying One Paradox in Twain's World

Note: Part 1 of this appendix will deal with simple statistical games. Part 2 will examine the math landscape of the Berkeley contradiction.

Part 1: The Gamesmanship of "Figures Don't Lie, Liars Figure"

America's reliance on statistics to mandate justice has created a brainwashed awe of...uh...Dammed Lies? Twain's antiquated advice has fallen on deaf ears for nearly half a century as modern statistics have gained validation in the courts and in the press. Statistical "bait-and-switch" and "comparing statistical apples to oranges" have simply overwhelmed our legal system. We'll briefly view how the statistics used to try to prevent unwanted discrimination are so easily hammered by Twain's, "Figures don't lie, liars figure."

Legislation enacted by Congress directs the judicial system to take reasonable measures to purge unfair discrimination from our country. Unfortunately, a semi-talented number-cruncher can review the hiring practices of any entity and garnish statistical facts that create issues. This will be seen in our typical example that represents thousands of ongoing sagas found throughout America today.

Let's say it is 1984 and you're a member of an oversight committee for a police department. You want to be vigilant that the police department is not being discriminatory. This year there are 100 openings for new police officers. The applications break down to 300 men and 15 women applying for these job openings. The police department fills the openings with 91 men and 9 women. Cross-town newspapers that always want to prove their competitor wrong are now ready for action. Opposing headlines scream: *"Blatant Sex Discrimination Continues: Less than ten percent of our new police officers are women!"* versus *"Our Mayor's*

Liberal Oversight Committee Goes Way Too Far: Police department favors women over men by over two to one!"

Neither headline is technically lying – less than ten percent of the newly hired were women...but...91/300 versus 9/15, means a woman applicant was twice as likely to get a job as was a man...and here come the lawsuits...from both men's *and* women's advocacy groups! Oh and oversight committee, shame on you. Twice!

So federal regulators tried to establish reasonable guidelines to lower the possibility that the courts are filled with lawsuits where **everyone** claims they were discriminated against. Here's what they came up with:

Basically an entity accused of discrimination is likely to be saddled with the major burden of proving its "unusual" hiring practices are fair if its hiring statistics fall outside of a "safe harbor" zone.[33] In other words, if an employer has hiring statistics within these safe guidelines, then the accuser has a very difficult case that is likely to be dismissed before trial. However, if hiring statistics do not fall within these guidelines, then its time for big legal headaches for an entity with deep pockets.

We'll look at just two of the more reasonable statistical guidelines to see how even the best of many guidelines can become perverted. Statistical guidelines can provide wonderful cover for those with personal vendettas and hidden agendas.

First Guideline: The 80% Rule

The 80% guideline establishes a policy that hiring practices are considered reasonably fair as long as at least 80% of applicants of a particular group's ratio are hired – if this happens then in all likelihood the judge announces, "Case dismissed" and everyone goes home to find something else to grumble about. So both men and women who felt they have been wronged are out of luck in our example above since the department would have to had hired less than 76 men or less than 4 women to fall out of the "safe zone" for this particular guideline.[34] And your oversight committee gets an official federal confirmation that it is doing its job.

---

[33] This "safe harbor" concept is common in the high risk investment community where typically the investors have signed that they have received packets describing the dangers of investing in a particular company....and if the investor is still dumb enough to invest, then he is usually out of luck in the courtroom when trying to re-capture any losses from a company that has made an effort to put itself into a "safe harbor" by providing these written warnings about investing.

[34] This is because $(100/315).8=76$ and $(15/315).8=4$.

Now comes the mischief. Let's say the mayor put Ms. Feminist on the police oversight board.[35] Her whole mission in life is to break women free from their slavery in a male dominated society and the mayor figures she is better suited for the police board rather than in his office pestering his staff so much. Her simple solution to these ridiculous guidelines is to go to the women's studies department at the local college and tell her enslaved sisters to all apply for the job in a self-righteous ruse so the total number of women that must be hired for the department to remain in the "safe" zone will be artificially high.

Now the police chief has had about enough. First of all, the Chief was always silently annoyed that the mayor had the audacity to even create an oversight committee reviewing *his department*...and then...saddle him with Ms. Feminist. So the Chief and the college's football coach decide to convince the whole football team to apply for positions on the police force too. The Chief and coach feel a true sense of accomplishment knowing their ruse will counteract the ruse from the "other side." And that's just the end of round 1 of this particular silly game that has been quietly going on for decades. The jaded create self-serving little cults of righteous indignation, and it's all relative to perceptions and past experiences as to who's righting a wrong and who is just being a jerk.

Bottom line: The figure pointing goes on and on and the workplace is demoralized as the focus is no longer on hiring the best person for a job, but more on cheating while appearing to follow an arbitrary employment policy maze that was corrupted long ago. The cynical play self-righteous games of vengeance masquerading as fairness...and lawyers representing those suffering from this massive injustice can start shopping for that expense vacation home they've always wanted – compliments of local taxpayers.

Second Guideline: A Pattern of Diversity

The courts in the 1980s started to notice how some victims of evil prejudice looked remarkably similar to earlier victims. The questions for someone who has sued the local police department, McDonalds, Starbucks and General Motors for racial discrimination are: Is racism really that prevalent and are all these entities really that evil? Or are you a bit like the three-year-old convinced that evil demons are nightly visitors to his bedroom? Or are you more calculating and just making a lucrative career suing everyone with deep pockets?

---

[35] Ms. Feminist will never be happy until the police force has close to 50% women police officers. Oh and the police chief silently harbors the ridiculous notion that men tend to be more suited to walk down dark alleys at 2 AM than are women. The stage is set for nothing but hidden agendas where everyone can feel cheated.

It's a believable scenario that a pivotal point in time occurred when an affable victim made so much money from settlements and jury awards that he moved into a luxurious mansion just down the street from the judge's house. And the judge then got a good dose of seeing what a jerk that affable victim can become when he's not on the witness stand displaying his best behavior. These sorts of events meant that further guidelines were needed to offer more insight into the statistical reality of discrimination...hmm. One of the more reasonable concepts was allowing a defendant to show a pattern of diversity in its hiring practices that would imply a mindset of overall fairness. So the saga continues as now "deep pockets" gather statistical data to show how many historically disadvantaged they actually employ. These statistics are gathered to try to convince a judge that the company belongs in the "safe harbor" and deserves a prompt "case dismissed."

So a woman employed in a job traditionally held primarily by men can be counted as a minority. A Hispanic woman has often been counted as two minorities rolled up into one person. Not that unreasonable of a concept until the police chief gets really cute. The Chief cackled himself to sleep the night he hired a Hispanic woman who was born with four toes instead of five on one foot and tends to have a nervous stammer when she is around the Chief. Let's see: 1) woman, 2) Hispanic, 3) physical AND 4) mental/speech handicap. Wow, four minorities wrapped into one, and she is the Chief's daughter-in-law! She can go on gravy-train disability with a wink from Dad when she wants to start a family. This fast track to her disability will improve still another noteworthy statistical study for the chief that tracks minority performance (or lack thereof). Her poor minority performance will prove how the Chief has bent over backwards to follow diversity guidelines that results in less talented police forced upon him. And who says America no longer embraces creativity!

Guidelines in General

A fundamental problem for statistical-guideline social solutions in theorem form is: For every new statistical guideline devised to fix a previous statistical guideline there probably exists at least two ways to circumvent the new statistical guideline's intent via mischief. Therefore, the flaws can grow faster than the remedies as questionable guidelines pile one on top of another to create one big guideline-statistical-legalized mass of mumbo jumbo. Perpetual second-guessing and round-table discussions can work hard to resolve one issue only to create more issues. A clever politician's dream comes true as each slighted group is quietly told, "I feel your pain and I'll make things right for you."

But how could even the most honest politician in the world "make things right"? Who really knows for sure whether the ratio hired versus the number hired is most relevant in our first example? How reasonable is the 80% rule? What would professional basketball teams look like if the 80% rule was mandated for basketball? More like the professional teams of the 1950s, except that about half of the white folks would be women? The black man in America would again be trampled over by other "minorities" as everyone jostles to get to the front of the preferential treatment line for a lucrative NBA salary. The 80% rule can clearly become a cause rather than a cure for discrimination. How often should statistical guidelines trump the concept of simply hiring the best person for the job?

There are also technical questions for our second guideline. Should a person with several historical handicaps be counted only in one category, or once in every category? And if counted only once, which categories should be artificially lacking? Which unusual characteristic of humans should be a category, and hence qualify for preferential treatment? Caldrons of questions with debatable answers go on and on.

Sooo, why not use what best suits your side of an argument and disregard the rest? Ergo: Use, "Lies, Dammed, Lies and Statistics" to your advantage.[36] Mastering this aspect of statistics has served many famous lawyers and politicians well. It can all boil down to "figures don't lie, good liars figure."

It is remarkable how years of college enlightenment have enlisted so many into cults that worship statistical guidelines as cure-alls for eliminating discrimination. Fortunately this brainwashing was never wasted. Naïve acceptance of statistical guidelines allowed the University of California (UC) regents to engage in truly philanthropic endeavors...such as ensuring that their offspring along with other UC policymakers could compete in the same arena with minorities and athletes for preferential acceptance into UC Berkeley and UCLA. After all, a regent's or professor's child is clearly ill-prepared to compete scholastically with the regular folks. Gee. Does this perk offer a little insight as to why so many professors favor a special minority quota arena? We'll let this bait-and-switch brilliance culminate our examples of "figures don't lie, liars figure." However, elitist mischief can distract us from our original Berkeley paradox. How can *every* individual department favor women while UC Berkeley as a whole favored men? There was no bait-and-switch, nor were statistical apples compared to statistical oranges since the same analysis was used throughout this study. Where's the mischief in these contradictory statistics? Berkeley's brainwashed encountered an impossible statistical illusion that is reality.

---

[36] The book, *How to Lie with Statistics,* is a great resource for those wanting to hone this skill.

Part 2: The Berkeley Paradox

First the most simple point: We can prevent these paradoxical results from occurring by simply keeping the group sizes the same. This conflict will never occur if each clinic had the exact same number of patients in each group.[37] One can get an intuitive sense of why this can happen when thinking of these fractions as vector addition.

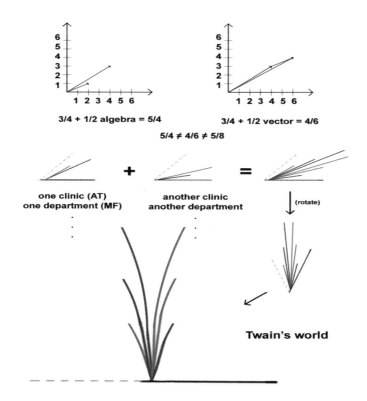

3/4 + 1/2 algebra = 5/4   3/4 + 1/2 vector = 4/6

5/4 ≠ 4/6 ≠ 5/8

one clinic (AT)   another clinic   (rotate)
one department (MF)   another department

Twain's world

Basically, the comparison occurs when class A vectors have a greater slope than their counter-part, class B vectors in every small group. The paradoxical result occurs when the small vectors are added and the resulting sum of vectors B has a greater slope than the sum of vectors A. This is not as uncommon as one would like to think. The thinking of just

---

[37] But this is not possible when extrapolating data from historical records. Even always creating new studies under this parameter of same group sizes may have occasional downsides because it is, in all likelihood, a distortion of reality.

keeping all group sizes the same is not possible when looking at historical data, and it can be easily argued that 'controlled' studies having same sized groups, while eliminating this particular paradox, can create still another set of distortions when evaluating human interactions.

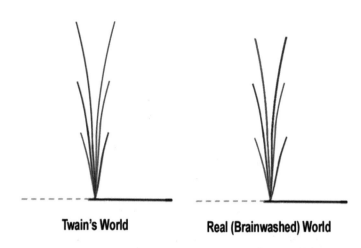

Twain's World   **Real (Brainwashed) World**

The continuing of the alternating of colors when two clinics are combined as above insures us that we have not entered into the paradox seen in our aspirin-Tylenol like comparison.

What set of ratios give us the greatest frequency of encountering Twain's World? Do some of the maximums and minimums converge to irrational numbers? If so, do we visit flavors of our old friends, π, or the natural log?[38] Clearly some Twain's World results are not counterintuitive; therefore, when do these results start to play tricks with intuition? This is a question for philosophers and psychologists. How would this affect the usefulness of the purely math convergence ratios discussed later in this appendix?

This paradox can be analyzed in a rather straightforward fashion. But the further we study this simple paradox – the more we'll find a detailed mathematical landscape that heads off in many different directions - as does Relativity. So this appendix is a great example for understanding the nature of detailed landscape surrounding the simple $E=mc^2$.

---

[38] Why can one be reasonably confident that many of these functions will converge to irrational numbers?

The flow chart basically starts counting all possibilities with vector ratios that do or do not fall into the paradox zone. This flow chart is an aid for showing the simplicity of the paradox and as a starting point for more useful programs and math analysis. The results are Rw for expected (i.e., the Real World for the brainwashed) and Tw is when the ratios (vectors flipping) enter into Twain's World.

The first flowchart can be divided into two parts; one being a looping mechanism for finding all possibilities to be tested and the second part containing comparison testing leading to other programs. A more insightful variation of the first chart is to focus programs with respect to vector sums rather than the vectors creating the sum.

A basic flow chart below is counting all possibilities of vectors 16 and under in the small groups:

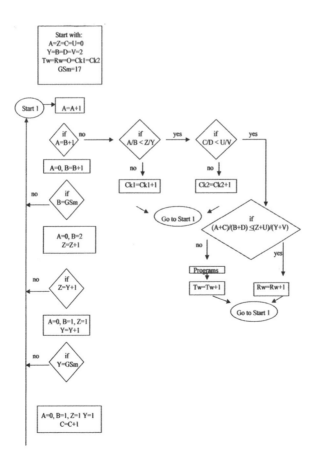

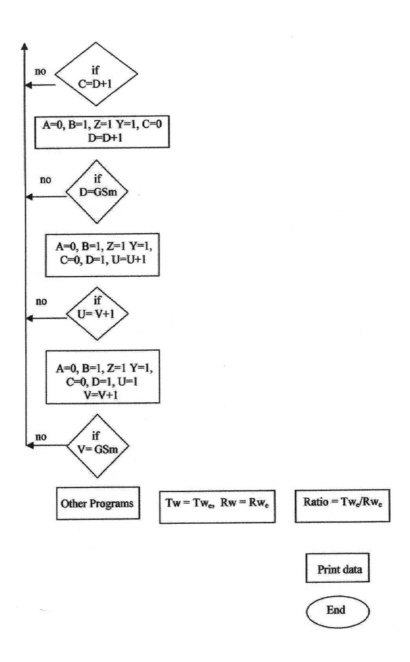

Now if we change the focus on vector sums (O,N,Q and P) then our flow chart will change. The original starts as:

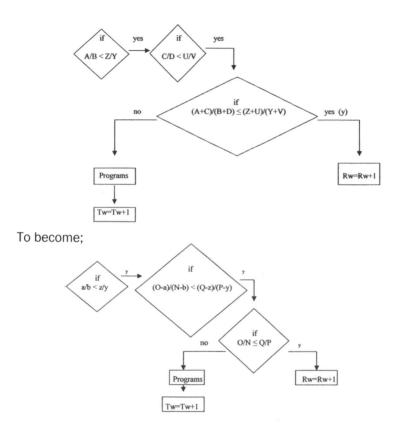

To become;

Now run every (a,b,z,y) within every O and Q, and every O and Q within each N and P

as integers Ns, Ps →∞.

If you have a big enough computer ☺

The looping mechanism in the original flowchart to find comparables is very simple and so helps as an intuitive aid of what's going on, and however, is needlessly over repetitive. It should be noted that slicker looping methods for the $O/N \leq Q/P$ format should be used with caution so as to not create problems similar to the probability shortcuts when not counting *all* events. In other words; taking logical shortcuts in the math world when exploring paradoxes, is inherently dangerous.

Since these results are vector locations in R^2, then as vector sizes change (increase) the results can be looked at as either a vector size change or a gridline change (shrink) where gridlines represent integers and vector slope flipping upon summation is our paradox. Therefore magnitude multiples will remain in the same zone to be surrounded by other dots (paired integers coordinates). This implies likely continuous paradox (backround) zones. Accordingly, the gridline shrinking or vector GT increasing should result in either rapid or eventual convergence to specific, likely irrational, ratio(s) is instinctively suggested - but not yet proven.

New nomenclature to study this further is useful. Therefore, starting from the flowcharts, we will define;

$$\mathcal{E}' \equiv Tw_e/Rw_e, \; \mathcal{E} \equiv Tw_e/(Rw_e+Tw_e),$$

$$GT \equiv N+P, \; \ni \; \equiv Rw_e/(Rw_e+Tw_e),$$

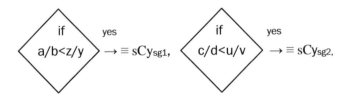

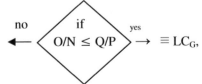

$_{GT}\mathcal{E}_{max}$ is the highest ratio for a given GT

Where sCysg1 and sCysg2 are small group comparisons that result in an accumulated group comparison LCG. Our original flow chart has just two small groups. The aspirin/Tylenol flowchart would have three small with one large comparison groups and would look like: sCysg1, sCysg2, sCysg3; LCG. And Berkeley having n graduate departments would have:

$$sCy_{sg1}, sCy_{sg2}, ... sCy_{sgn}; LC_G.$$

Nsgn ≡ the number of small groups compared.

Defining € further, for Nsgn = 2, $_{GT}€_{max}$ has a degree of ambiguity[39] with corresponding O_{max}, N_{max}, Q_{max}, P_{max}, and another layer of O'_{max}, N'_{max}, Q'_{max}, P'_{max} with *fixed* b'_{max}, y'_{max}. We'll define $_{GT}€_{max}$ as the larger of the two. How often does $O_{max} = O'_{max}$ for a given GT? Also, how does Δ $(N'_{max} - O'_{max})$ compare with $Δ(N_{max} - O_{max})$?

This should give us enough new nomenclature to describe useful functions.

Another set of simple examples:

If for a GT, an N, where $N'_{max} ≠ N ≠ N_{max}$ then, N would have a set of b_{max}, y_{max} for its $€_{max}$. Therefore as the N travels its region; GT→N→0 (and its P goes; 0→P→ GT) there are variations of b_{max}, y_{max} that can be described as a non continuous function in a continuous backround of Tw_e or Rw_e for the N-P pairing of a particular GT. As GT→∞, what do O_{max}/N_{max}, Q_{max}/P_{max} converge to?

€ has layers of detail in multi-variate directions.
(Note: Nsgn is assumed to be two unless otherwise stated.)

We'll now define broad multidimensional functions whose inner workings often contain a useful class of functions defined by; $_{GT}€→§$ as GT→∞,

A noteworthy concise example is;

$$(_{GT}€_{max}→ §'_{max} \text{ as } GT→ ∞) ≡ \int_{(GT}€_{max}).$$

What is the specific description of §'max at ∞?

What is the function, $\int$, for the variation of € for a particular N-O localized wandering look like? (On a macro scale, are their convergences

---

[39] Uncertainty and ambiguity until more detailed nomenclature is defined.

115

harmonious or shades of chaotic convergence like Feigenbaum's program of, $x_{n+1} = 1 - ux_n$?[40])

How does $\int(\text{€}_{max})$ change as Nsgn→∞?

A down-to-earth question that is very solvable in this wilderness is:

What's the highest probability (likelihood) of all the different graduate departments – by chance – wandering together into Tw as was done in our real-life example? Let's put the number of UCB graduate departments at 10 (Nsgn) and total applications to graduate schools at 5,000 (GT).[41] We now have enough information to find an answer to our question. We would be looking for a specific $_{Nsgn-GT}\text{€}_{max}$. **What number is $_{10-5000}\text{€}_{max}$?**

What are (($_{10-5000}\text{€}'_{max}$),

(($\frac{1}{2}$)^{10}($_{10-5000}\text{€}_{max}$)/($_{10-5000}Ә_{max}$)), and ($\frac{1}{2}$)^{10}?

Which of the above is the most accurate cap answer to our UCB question?

What formulas would be much more accurate caps?

(How does it compare to P<.05, or the chance of dealing four of a kind in poker? Golly gee...whatever does that say?☺)

Also what about the class of (Nsgn – 1) like dilemmas mentioned on page 11?

So we have several $\int$ of reasonable interest;

some being $\int(\text{§}'_{max})$, $\int(_{Nsgn-GT}\text{€}_{max})$,

$\int(\text{§}_{max-Nsgn})$ c ($\int(\text{€}_{n-o})$u$\int(\text{€}_{Nsg})$)

(The last combo being a monster in size.)

When it is all said and done, what are the important landmarks on this map of paradox?

---

[40] See appendix E for more on simple chaos.

[41] Note: These are not the actual numbers of Berkeley's 1970 admission data, but just convenient estimates for a nomenclature example resulting in a specific $_{Nsg-GT}\text{€}_{max}$ that has not yet been calculated.

Appendix B

Binomial QM Spikes and Otherwise

A simple example to complications of statistics and quantum mechanics is a simple variation of dice. Change the 3 to a 2 and the 4 to a 5. With 2 rolls of the die, the single peak becomes three and interestingly the frequency of obtaining 7 increases.

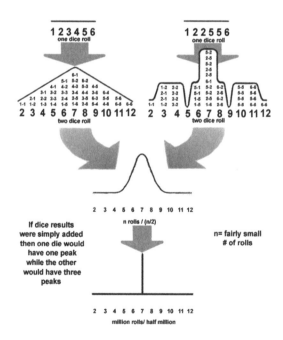

The problem of knowing what we are observing becomes important. Are we observing accumulations of successive events or some form of averaging events out? The later starts to blend the two very different underlying events as one of the same.

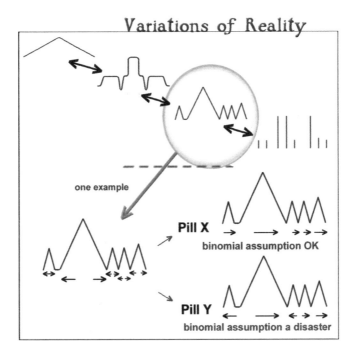

Variations of Reality

one example

Pill X
binomial assumption OK

Pill Y
binomial assumption a disaster

Every genetic marker can be thought of as a possible peak with "whatever" affecting those with that gene. How many genetic peaks and which ones cause, or is at least a marker, to a difference in therapy response? A clinical trial to "cover all the bases" is almost always impractical. Today's thorough clinical parameters that would try to include genetic variations will likely require group sizes to be larger than all those in the USA inflicted with the malady!

Quantum mechanics has a greater set of issues. One question is how many quantum events are like the cosmology questions with respect to stars are very observable but "dark matter" is not. How much does the difficulty of dark matter evaluation contribute to our unknowns in stellar observations? In the quantum world: how many of the peaks are observable and how many are not? Are the peaks accumulations or some type of median/averaging? And how do our observations deceive us regarding interactions? Where's the action and where's the location?

A good illustrative example is the swing of a simple pendulum. It always slows as it changes from going from one direction to the opposite direction. The ball will spend much more of its time changing direction than speeding along the bottom of its swing. Yet the pendulum will deliver the most 'punch' with an interaction collision at the bottom of the swing when the ball has its greatest kinetic energy. So if one observer simply uses a camera that randomly flashes and notes the location of the ball and another observer only notes when the ball's reaction to collisions is at its zenith, then the two observers will have two very paradoxical observations where mathematical creativity is useful in suggesting many logical possible solutions to harmonize these observational conflicts. Some logical suggestions will describe reality, while other logical suggestions intellectually solve the apparent paradox – but are not relevant to any other aspect of the pendulum. This illustrates the many joys and challenges in quantum mechanics. The layers of intrigue of "what's going on and where," are enormous because of probability, statistics and our lack of complete observation.

Appendix C

Multivariate Evaluation

Unfortunately, the creation of resistant bacteria was needed to expose the flaw with a "one active ingredient study approach" for therapy, to accommodate clearer statistical results. Agra-business has had to deal with this problem for decades earlier than did medicine. It is a pity that more interdisciplinary interaction has not occurred. The medical world for humans could have been easily forewarned of this problem by farmers.

Below is a simple pragmatic way to evaluate several different active ingredients and look for both negative and synergistic interactions. First find the appropriate starting point from a dosage/response curve for each active ingredient (ai). The optimal evaluation of the first four variables will be determined and then will be combined with the next three variable active ingredients (ais) and the process will repeat itself. The example assumes there exists a good standard bioassay plate in standard 96 plate two dimensional matrix form. The concept can be extended to other types of bioassays. Verification from one assay can be evaluated by another assay and to eventual animal studies. Human trials for tweaking are likely too expensive – clinical evaluation would typically be only with the best multivariate bioassay results compared to a single agent (ai) and the control.

This method can also be used to gain insight on more subtle interactions. Active ingredients, ais, can be replaced by active environments, aes, (lipids, protein-peptide-aminoacid, polyphenols, citric acids, terpenes, and other nutritionals) in this multivariate model to evaluate gastric, plasma conditions, and effects on efficacy of ais. One should not anticipate reaching statistical significance (P<.05), but rather, variations of averages when conducting animal trials for aes.

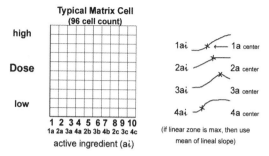

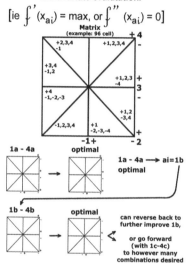

Such overall results could offer suggestions for avoiding negative Rx interactions and enhancing possible complimentary benefits including, which foods are best consumed with different medications. The degree to which diet interacts with medication is still poorly understood in this country. This is partly because in almost all cases, there is likely to be only a modest change, if any, for each particular ai. But as our population ages, the number of people taking several different types of therapy at the same time will increase dramatically. One modest therapy variation added to another over and over again could result with the accumulated modest variations becoming a dramatic improvement, or alternative, a dangerous combination.

As an example: Let's look at a particular cancer evaluation option of an injection of a cocktail into an organ where both the cancerous and normal cells are exposed to the medication. A 96 plate containing equal amounts of both cancerous and normal cultured cells receives chemotherapy agents. Each of the 96 cells on the plate will have varying quantities of 4 different active ingredients. 1a, 2a, 3a, 4a would be cisplatin, 5-FU, cyclophosamide, and a uracil-catechin complex respectively. A dose/response curve would be established for all four chemo agents. Cisplatin would be axis 1 (vertical axis) where increasing dosage occurs in the right direction to the axis while a decreasing dosage occurs to the left. Then middle cells would have ¼ dose of the middle of the previously established dose/response curve and an appropriate increase/decrease would continue to the exterior cells of the matrix plate. These exterior doses will usually start at a low end being (left side) below the bottom of the curve; to the other extreme (right side) being three-quarters up the dose/response curve's range. Say in this example that cisplatin had a dose/response curve starting at 5ng/l and the curve flattens again at 60ng/l. Then the middle cell columns would be; 7ng/l in the left center cell column and 10ng/l on the right center column (about ¼ of middle of d/r curve). The far left end cell column would have 1.5ng/l and the far right column row would have 45ng/l. (In other words, the right side of axis would be 10, 16, 24, 33, 45 ng/l; while the left side of axis would be 7, 5, 4, 3, 1.5 ng/l.)

The next ai, 5-FU, would run along the first diagonal axis and would have 20 variations of dosages (because of the checker moves for the diagonal axis). These dosages would be established in the same manner as was cisplatin. Next would be horizontal axis of cyclophosamide, which would be calculated like cisplatin, and the changes of dosages in cell rows would run perpendicular to cisplatin's columns. Finally, the uracil-catechin would run perpendicular to 5-FU and on a 45-degree axis to cisplatin and cyclophosamide.

The sweet spot in the matrix would be the area of adequate cancer cell deaths accompanied by minimal healthy cell deaths. This particular concentration of cisplatin, 5-FU, cyclophosamide, and uracil-catechin concentration would then be 1b, and 2b could be gleevec (imatinib) and 3b and 4b could be taxol derivatives (paclitaxel and docetaxel) respectively. The same cell lines could evaluate this combination. (In other words, a total of seven active drugs would then be evaluated in a manner as were the first four agents. Four are more fixed than the last three drugs.) This optimal combo would then be 1C. The last of the ten (2C, 3C and 4C) could be marine chemo-agents spongistatin, auristatin and gemcitabine respectively to round out the third multi ai combo-plate. This would evaluate ten active agents for a particular cancer cell line.

Further runs could either: tweak optimization by fine-tune existing ais; or add additional active ingredients or aes (such as, active environment factors that potently activate inflammation modulators, or simpler environmental factors such as lipids terpenes, etc.).

Appendix D

Escape Velocity Calculation

(Note: One needs to have a modest comfort with calculus for some of the math in this appendix)

This appendix will calculate how fast an object must be leaving our earth so that earth's gravity will not drag it back down. The calculations will assume no air resistance, nor other celestial bodies, nor other minor factors interfering with the accuracy of the forthcoming equations. Therefore this is just a rough estimate of the actual escape velocity, but it successfully counters the logic to the old axiom, "What goes up, *must* come down."

The escape velocity can be calculated by assuming the following; a mass no matter how far away from earth, with the smallest of motion toward earth, is attracted to earth by the gravitational equation;

$$GM_em/R^2$$

where G is the gravitation constant, M_e is the mass of the earth, m is the mass of the object infinitely away from earth and R is the distance between the two masses from their centers.[42]

Many observations and experiments (in classical physics) have shown that the increased energy of an object moving closer to earth because of gravitational attraction is the integral of the force above;

$$\int GM_em/R^2dr$$

---

[42] Gravitational Constant = $G = 6.67 \times 10^{-11}$ nt-m/kg, Mass of earth = $M_e = 5.98 \times 10^{24}$ kg, radius of earth = $r_e = 6.37 \times 10^6$ meters.

where the integral's boundaries are initial/final locations, which in this case would be infinity (initial), and the radius of the earth, r_e (final).

This integral is solved to become;

$$GM_em/\infty - GM_em/r_e \;\longrightarrow\; GM_em/r_e = \text{kinetic energy of mass striking the earth.}$$

This is then also equal to;

$$\tfrac{1}{2}\,mv^2$$

which is another description of kinetic energy, therefore;

$$GM_em/r_e = \tfrac{1}{2}\,mv^2 \longrightarrow GM_e/r_e = \tfrac{1}{2}\,v^2$$

Now a simple trick, if the speed can be calculated for this mass striking the earth, isn't this the same as the speed needed at earth to send the mass an infinite distance away from earth?

If so, then we can solve for v because G, M_e and r_e are all known.
$$GM_e/r_e = \tfrac{1}{2}\,v^2 \longrightarrow (66.7\text{X}10^{-11})(5.98\text{X}10^{24})/(6.37\text{X}10^6) = \tfrac{1}{2}\,v^2$$

This is solved to v being approximately 7 miles per second (or 11.2 km/sec) – the escape velocity for earth's surface.

Ergo – "What goes up, _must_ come down," is definitely wrong.

More Fun with classical physics and $GmM_e/r_e = \tfrac{1}{2}mv^2$ if basic relativity (Chapter 3) is understood. It will become apparent why a daydreamer would so easily abandon the conventional wisdom of 1904.

The equations $\int GM_em/R^2dr$ and $GmM_e/r_e = \tfrac{1}{2}\,mv^2$ accurately predicted the earth's escape velocity in a very slick manner and is also great in showing other simple paradoxes in _inaccurate_ classical physics, where some of these paradoxes are solved, while others become more confounding when extended to Relativity.

If $GM_e/r_e = \frac{1}{2} v^2$ tells us that 7 miles/second is the escape velocity for our earth, then how dense would the earth have to be for $v = 186,000$ miles/second escape velocity? The classical physics answer would be that the earth must be about 660,000,000 times denser for the mass to strike the earth at 186,000 miles/second. Now what if the earth was even 1,000,000,000 times heavier still, how fast would the mass strike the surface of our very heavy earth? About 300,000 miles/second? Nope – no faster than 186,000 miles/second since that is the speed of light, c, and nothing goes faster than c. Otherwise our *inaccurate* classical method would eventually have the object's kinetic energy much larger than $E = mc^2$ when striking the earth; *or* the mass reaching its maximum speed, c, way before it reached the earth because of $GM_e/r_e = \frac{1}{2} v^2$, and because nothing travels faster than the speed of light. The characteristics of the c boundary within Relativity keeps today's conventional wisdom moving smoothly. The concept of gravity and electromagnetics being fairly equivalent answer questions like; do gravitational objects continue their classical acceleration to 186,000 miles per second and then instantaneously stop accelerating? [43] Or is gravity like electro-magnetics and has a curving of mass/space as particles start to reach c? How similar is gravity to electromagnetics? We can derive Relativity from experiments and Maxwell's equations – but in classical physics we have a maximum kinetic energy of $\frac{1}{2}mv^2$ becoming $\frac{1}{2}mc^2$. What happened? Where's the other half of $e=mc^2$? Tons of other questions abound like: When do black holes occur? And since our trick in Chapter 3 says light always travels at c no matter what – is the trapped light circling within black holes like our moon rotating around earth or is there a point when even light collapses into an abyss? Can a black hole that has the same mass of earth exist at about 1/4 inch thick, or do black holes collapse into singularity? And how do Chapters 4 and 5 hinder arguments of singularity?

Another simple paradox in classical physics is, if we just look at the GM/r part of the equation $GmM/r = \frac{1}{2} mv^2$, and note that as $r \rightarrow 0$, then $GM/r \rightarrow \infty$, so also $\frac{1}{2} v^2 \rightarrow \infty$...right?

---

[43] Note: This gives a flavor of why the reasonable assumption of equivalent behavior of masses with respect to gravitational and electromagnetic forces was made. These figures and thoughts are just convenient fantasies used as a simple method to get a sense of the possible direction of relativity and where a more complicated reality beyond relativistic equations might go.

But if we look at $\frac{1}{2} v^2 = \frac{1}{2} d^2/t^2$, we know that d is limited – *capped*. In other words, $r \rightarrow 0$ implies $d^2 \rightarrow \infty$, but d is capped $=><=$.

Also piling on is; as $r \rightarrow 0$, then $t^2 \rightarrow p>0$ but generally, x^2 overrides a simple y within the world of convergence/divergence limits.

Both d capped and how $t^2 \rightarrow p>0$; make the logic of;
$$r \rightarrow 0, \text{ therefore } GM/r \rightarrow \infty$$
very questionable in terms of relevance to reality.

But this thinking still leaves us dangling for answers and in a world of uncertainty in classical physics.

In other words, there are several ways that classical physics, by itself, tells us there comes a time when,

$$\int GMm/R^2 dr = \frac{1}{2} mv^2, \text{ and related equations}$$

break down as being relevant. One can easily see why Einstein was quick to abandon classical physics with the benefit of a century of **hindsight.** Einstein's assumption of Cem and that c is the max for v, are practical solutions for many theoretical dilemmas. But are these "solutions" appropriate for all occasions?

Examples of "Proof by Contradiction" and Mathematical Chaos

This appendix will give a brief flavor of "proof by contradiction" and chaos theory, by showing a few examples and explaining how they apply to our main presentation in this book.

Use of Contradictions in Math Proofs

Proof by contradiction is usually:

Is something, we'll call 'omega', true?

We'll assume omega is *not* true.

Then with omega not being true, equations are constructed and manipulated until one comes to an obviously false conclusion (example, often zero does not equal zero).

Therefore omega must be assumed as possibly being true since the converse is not always true. QED

(Note: There is reasonable debate that *some* contradictory proofs have subtle flaws.)

The Classic "Proof by Contradiction" Example

The most famous 'proof by contradiction' is showing how $(2)^{1/2}$ must be an irrational number.

We will first describe in pure prose what we will do in this proof, followed by a formal proof. In prose we will describe our assumption, the contradiction, <u>and</u> our only logical alternative.

Pure Prose Overview:

We will first *assume* that radical two is a rational number.

We will first *determine* that an
<u>x and y **cannot both be even**</u> if radical two is a rational number. *Yet later on we will <u>determine the same x and y **must both be even**</u>* if radical two is a rational number...well it can't be both!
...<u>*an obvious contradiction*</u>...

Therefore our original <u>*assumption*</u> that $(2)^{1/2}$ can be described as a rational number must be wrong.

Therefore $(2)^{1/2}$ must be the alternative, which is an irrational number.
QED (end of proof)

Formal Proof that $(2)^{1/2}$ is an Irrational Number

Let's *assume* the opposite is true – in other words, that $(2)^{1/2}$ is rational. Then $(2)^{1/2}$ can be described as $(2)^{1/2}$ = x/y where x and y are integers (the definition of a rational number).

Now let's make our x/y the set of integers 'reduced to lowest terms' (rtlt) of all integer ratios that equal $(2)^{1/2}$. Therefore x and y cannot have a same factor by definition. (i.e. no common factors exist in both numerator and denominator because x/y $\equiv$ rtlt of $(2)^{1/2}$).

Therefore 2 can be a factor in either x or y but not be a factor in both x and y, or 2 is not a factor in x nor y. In other words, rtlt mandates that **either x or y *must* be an odd integer.**

We now have enough information to show a contradiction.

$$(2)^{1/2} = x/y$$

$$2 = x^2/y^2$$

$2y^2=x^2$, therefore x^2 is an even number, hence x must have a factor of 2, otherwise x^2 would be an odd number. Therefore, **x must be even.**

Since we are dealing with rtlt of $(2)^{1/2}$ we have previously shown that y *cannot* also be even.

Since x is even with a factor 2, we'll define x $\equiv$ 2p.

Therefore $2y^2=x^2= (2p)^2$

$2y^2 = y^2 + y^2 = (2p)^2 = 2^2p^2 = 4p^2 \rightarrow y^2 + y^2 = 4p^2 \rightarrow y^2 = 2p^2 \rightarrow 2p^2= y^2 \rightarrow$ y *must* also be even as was just shown above for x.

Therefore **both x and y *must* be even.** **=><=** (this symbol denotes a **contradiction has just occurred** and is often used in these proofs as their conclusion.) Therefore $(2)^{1/2}$ is an irrational number.

QED*

Generally, proof by contradiction shows that an underlying converse assumption cannot always be assumed – or as in this case – is not true. When using the weaker form of proof by contradiction (i.e., the assumption is not *always* true and whether you're dealing with a weaker form or a more definitive form is not always easy to distinguish), one is left in a gray area of uncertainty. In other words, one can't always rely on the particular assumption. Wariness becomes a risk adverse de facto assumption – if you can't rely on the assumption all the time then why rely on it at all? Especially if you're a PhD candidate that has to come up with a thesis that will pass the scrutiny of your mentors. Worse yet, you might spend endless hours assuming the assumption, one way or the other, when in fact the alternative would serve you best (never let it be said math is always easy). These facts did not discourage Einstein. One can argue that continuing beyond contradictions can easily put physicists into complexities which makes the reading of 'Alice in Wonderland' very entertaining.

Einstein "fixed" one contradiction by proclaiming something has to give in Special Relativity...and making it time and space. This matched experimental, real-life equations made by Lorentz. Maxwell's four equations came to be because they solved real-life problems in the field of electricity within the confines of Cem. $E=mc^2$ can be easily derived directly from the Maxwell equations. Einstein was clearly aware of this fact well before 1916.

Bottom line, there are two separate ways to derive $E=mc^2$:

1) Directly from the practical 1873 Maxwell equations validated by describing the reality of electricity, and

2) Einstein's 1916 proclamations of a new Cem, which are solutions to contradictions.

The practical confirms the theoretical solution to contradictions for Einstein.

The popular concept in sci-fi of parallel universes also comes from mandating Cem as a particle's speed comes close to c. Its mass appears enormous in terms of increasing its speed and the parallel universe is either reality, or a very sophisticated allusion to accommodate Cem. A good example of this is: $V = V = V'$ to accommodate conservation of momentum in our Example 2 to derive $E=mc^2$. Might solutions like parallel universes and strings be more similar to Cicero's success than first glance would show? Cicero's solution to political uncertainty is not really a **solution** to a paradox but rather an incredibly useful **accommodation** to a paradox for the benefit of mankind. How different is this to our limited validations of Relativity? Is a lack of progress occurring because our knowledge of mathematics is too limited and/or we have made too many overly restrictive assumptions?

(*Side note: The flavor of the month of America's "politically correct" mathematical education is to encourage prose to replace 'older' proofs using universal symbols lacking proper English sentence structure. Is this a wise policy to advance mathematics and science? Try to write the simple formal proof on page 129 in proper sentence structure. If the effort is made, then present the prose to others and evaluate the difficulty to comprehend such simple logic in only prose form. The speed and parallel logic regions of the brain are needlessly distracted and placed in neutral as time and energy is dedicated to fitting slick logic into a clumsy ancient esoteric form of communication for logic – the proper sentence structure and grammar of English. Hidden agendas to eliminate the talent of some to appease intellectual vanity of others should not prevail in guiding our educational policies. There are those in education who are inclined to recite, "Mirror, mirror on the wall, am I as smart as the smartest of them all?"... and then proceed to fix things. Education should acknowledge that such vain people exist everywhere and resist urges of denying others the chance to excel at their unique talents in the name of fairness. This applies to all human endeavors – not just theoretical math. Unusual methods to express slick math logic should be encouraged – not suppressed. The whole goal is to convey logic succinctly! Did Twain and Einstein enjoy crafting aphorisms partly because they say so much with so little?

This lesson of communicating logic goes even further. Did the uneducated Michael Faraday make remarkable theoretical breakthroughs partly because he lacked the knowledge, hence the burden, of slick math-speak-ease that happened to be still too cumbersome for his type of cerebral logic? ☺)

Chaos Examples

Chaos theory is another interesting area of study. We'll briefly touch on prime numbers, orbital trajectories and simple computer generated models. These simple chaotic models will then be compared to a pill entering the human body.

The chaos in prime numbers

The sequence of prime numbers is a good simple example of chaos theory and it accompanying debate. When will you encounter the next prime number is a mystery that can currently only be solved by methodically checking successive numbers. Mathematicians like Karl Mahlburg and Bruce Berndt keep whittling away at reducing our level of uncertainty of which successive number might be prime. But there is still no known way to predict the next prime number. In other words, chaos exists when just looking for a simple characteristic of simple integers. But is this reality or will we become smart enough to predict the next prime in some slick manner (or is this answer already known to the inner bowels of the CIA and naval intelligence)? Can today's chaos of prime numbers be solved to become tomorrow's exactness? Determinists have this goal for all that is called chaos.

Poincare's favorite chaos

Many three dimensional three body problems looked at by the famous French mathematician Henri Poincare are incredibly chaotic. This is very different than our solar system. Its future can be calculated with incredibly accurate detail. Our roughly two dimensional solar system contains well over ten bodies rotating around a stabilizing sun. Poincare's work was done over 100 years ago when determinists did not have the argument of computers outperforming human number crunching capabilities. Many of Poincare's once chaotic three body examples can now be made accurate to an incredible level with our most powerful computers – others are still out of our computing reach. Also, one can still argue for chaos by extrapolating from the following argument found in a simple billiard ball example. Its implications are far reaching:

1) You can never have a cue ball hit another ball dead center. There is an apparent 'Y' in this action -- the ball will either veer slightly to the right or to the left. *If* going right creates a cascading of events S, while going left creates a **very different cascading** of event W, **and if** one lacks the ability to know which way the ball will go off center, **then** chaos can still exist. [44]

Maybe billiards' popularity comes from the instinctive pride good players have to control seeming chaos into exactness as balls are accurately sunk into pockets. But even the world's best pool player can control the chaos on the table only so far.

Another example of this type of cascading "Y' is the free $5 coupon given to a hotel guest at many casinos. A cautious guest might figure the following no lose option - use the free coupon to make one free $5 play at blackjack but only continue to play with money made from accumulated winnings from his first free bet. The "y" is the result of his

---

[44] Determinists' counterargument for squashing this theme is we don't need exactness; only a slightly greater level of exactness to the uncertainty we've encountered to prevent unknowns of opposite cascading.

first bet. A loss means the early start of a nice long sound slumber in a hotel room (event S). A win might cascade into an exciting night of winning big (event W). Two very different nights determined in the brief seconds that it takes to flip few cards. This concept of chaos is nothing new:

"For want of a nail, a shoe was lost;

for want of a shoe, a horse was lost;

for want of a horse, a knight was lost;

for want of a knight, a battle was lost;

for want of a battle, a kingdom was lost;

all for want of a nail."

- George Herbert, early 1600s

2) Once you accept that 1 is possible with pool balls then our existence is filled with other subtle Ys.

The simplest chaotic Poincare three body problem is three equal mass spheres where two are in perfect circular orbit and the other trapped in prefect perpendicular oscillation about the middle of their axis. Assuming an external variation surrounding these objects, then an eventual variation from perpendicular will result into a complicated chaotic cascade. Also, the more exact the circular, perpendicular and subtle the surroundings, then the more difficult it is to predict the eventual collapse of circular-perpendicularity harmony – a more precise to almost exactness exacerbates the unpredictability of eventual chaos!

Computer Programming

Another good example of chaos is M.J. Feigenbaum's computer generated programs that have areas of a sense of randomness. The most famous is a simple sequencing of;

$$x_{n+1} = 1 - ux_n$$

and u is a number between 0 and 2. It has regions of reasonable convergence to exactness and other areas of apparent chaotic diversions.

His programs have similarities to many of Poincare's three body orbiting problem.[45]

The math crowd ranging from PhD candidates to the tenured in the 1970's were encouraged to examine how just a few simple continuous functions combined can create numerous unknowns that enter into the worlds of both uncertainty and chaos. There are **many, many** other well studied examples of chaos.

Chaos in Medicine and Its Denial

The above scenarios are all incredibly simple when compared to the human body taking a pill and predicting a definitive outcome. Chaos can occur with oh so few simple variables. The human body has how many complexities each containing how many unpredictable and unique variables for each person? When should therapy be denied to all if there is a very strong but chaotic trend toward therapeutic benefit accompanied by a few chaotic bad exceptions? Now a few easy questions: Which is **more** chaotic – our chaotic medical statistics or our chaotic judicial system? Which would you prefer to decide if a therapy is available for you – a semi-random jury of your "peers," or medical experts reviewing expensive albeit chaotic medical statistics?

Our legal tort system has the "victim" in plain view of a compassionate jury to the joy of PI attorneys. The benefactors of medical therapy are not to be seen as they simply continue with their improved lives. Worse yet, **hindsight** can easily paint the plaintiff as sinister for failing to predict the unpredictable.[46] This type of legal presentation to a jury is inherently biased, has way too much chaos, too many shams and unrealistic expectations to be **the de facto** arbitrator for future medical options. Left unchecked, it will deny us all many somewhat risky – but still by-in-large good medical options – particularly for debilitating diseases like MS.

---

[45] For further reading: Exploring Chaos, edited by Nina Hall, Mathematics and the Unexpected, by Ivar Ekeland, and Gotcha, by Martin Gardner.

[46] This human trait of the next-day potatocouch-coach obviously knowing better *after* the game what the high paying pros in sports *should* have known *before* the game applies even more to medicine. The prevailing factors of so many unknown variables become oh so clear to the juror who only deals with *hindsight.*

> "If I have a thousand ideas and only one turns out to be good,
> I am satisfied."
>
> *- Alfred Bernhard Nobel (1833 – 1896)*
> *Swedish chemist and benefactor of the Nobel Prize*

Prelude to Practical Tabooed Physics
(Appendices F and G)

A respected Cal Tech group headed by Jeff Kimble has received lots of funding in its efforts to use "spooky action at a distance" for the Internet's benefit. The Internet is essentially computers, routers and repeaters. Cal Tech is trying to create repeaters that could work much faster than the speed of light via light polarization. This light entanglement repeater would appear to be instantaneous by our methods of determining the speed of things, and of course is in direct contradiction with Einstein's theories that assume nothing travels faster than the speed of light.

Now, if repeaters might possibly run almost instantaneously, why is it impossible that switches cannot be made to run at spooky action speed? Once a spooky clock/compiler exists, then tomorrow's computers would make today's fastest computer look like an old abacus! And compared to the speed of human synapses...well, we'll let Michael Crichton and James Cameron go there.

Appendices F & G are attempting to show the feasibility of other devices considered by most as impossible. They don't challenge the assumed mantras of fundamental Relativity, but rather challenge the equally entrenched conventional wisdom regarding the Second Law of Thermodynamics.

One needs to be comfortable with higher mathematics, the Maxwell equations, and physics in general to understand what appendices F and G are trying to describe. They claim to be somewhat analogous to the sipping bird.

Opposite warnings regarding these speculative appendices are:

1) Sponge machines don't work despite reasonable logic that there exists an imbalance of forces if friction can be overcome. Inventors of perpetual motion machines usually fail to account for some force that actually keeps the system in balance. This neglected force is often very

difficult to find. So this phantom imbalance is, in fact, balance with equal and opposite forces...and the unfortunate inventor's machine will never work.

2) Equal and opposite forces were strongly argued for why a particularly "crazy" machine would never work. It was unanimous in academia that this machine was a real loser. The naysayers' logic actually sounded fairly reasonable – almost as sound as, "what goes up, must come down." The inventor of this crazy machine has been caught and preserved beautifully on film. He colorfully describes how all in academia thought he was really stupid. His facial expressions and descriptions of naysayers' reasoning for why his crazy machine wouldn't work are the ultimate examples of a human looking like "the cat that caught the canary." He was the inventor of the jet engine!

Second Law Violations in the Wake of the Electrocaloric Effect in Liquid Dielectrics

By Andreas Trupp

Appendix F follows on pages 138-153. Reprinted by permission of the author.

Second Law violations in the wake of the Electrocaloric Effect in liquid dielectrics

Andreas Trupp, Fachhochschule der Polizei des Landes Brandenburg
-University of Applied Science-
Private e-mail: atrupp@aol.com

A short version of this article was published in:

AIP Conference Proceedings Vol. 643: Quantum Limits to the Second Law, First International Conference on Quantum Limits to the Second Law, University of San Diego, California, 28-31 July 2002

Abstract: In any textbook on physics, Coulomb's law of the mutual force between two point charges q_0 at a distance r is modified by the appearance of the term K if the point charges are embedded in a dielectric:

$$F = \frac{1}{4\pi\epsilon_0} \frac{q_0^2}{K\, r^2} = 4\pi\epsilon_0\, K\, V^2$$

The dimensionless constant K (≥ 1) denotes the permittivity of the dielectric. According to this formula, the force is either *reduced* by the factor 1/K -if the charges q_0 are kept invariant in amount-, or is *increased* by the factor K -if the *potential* V is kept invariant- as a result of the introduction of the dielectric. V is the potential of the location of *one* point charge as a result of the field generated by the *other* point charge. Feynman argues that the formula is correct only if the dielectric is a liquid, and that it does not work properly with solids. His criticism does not go far enough. Two simple experiments with a liquid dielectric (backed by theoretical reflections) reveal that the formula is correct only if the two point charges have opposite signs (negative and positive). If the signs are equal, the formula reads (when applied to point charges in liquid dielectrics):

$$F = \frac{1}{4\pi\epsilon_0} \frac{q_0^2}{K^2\, r^2} = 4\pi\epsilon_0\, V^2$$

Hence the force is reduced by the factor $1/K^2$ if the *charges* (of equal sign) are kept invariant, and is left unaffected by the introduction of the dielectric if the *potential* V is kept invariant. With a so revised formula, cyclic processes can be performed in which the electrocaloric effect (that heats up the dielectric when the electric field is being built, and cools down the dielectric when the field is disappearing) is no longer symmetrical, leading to the conversion of ambient heat to electric work as a net result of the work cycle.

1. Qualitative description of the Second Law violation

It is a well established fact that internal energy in the form of heat, taken from a single reservoir (the ambient) only, can be completely converted to mechanical work without any refuse heat (that would be an unwanted side-product). Such a conversion is, for instance, taking place when an ideal gas, concealed in a cylinder, is expanding isothermally at ambient temperature and is thereby moving a piston. The only heat flow involved is that from the ambient into the gas (to prevent its cooling off), with that heat being totally turned into mechanical work performed by the piston.

Unfortunately, a perpetual motion machine of the second kind (that would convert ambient heat to work without requiring a second heat reservoir of a lower temperature for the reception of refuse heat) can nonetheless not be created thereby. For such a machine to operate, a **cyclic** process would have to be performed. Any means of getting back to the starting point (that of the compressed gas in the cylinder at ambient temperature) would, however, be accompanied by a creation of heat received by the ambient. Moreover, the quantity of heat thus delivered to the ambient would -at least- be as large as the amount of heat that was previously taken from the ambient. With no heat taken from the ambient as a net result of the cycle, no net mechanical work is yielded as a result of the cycle (otherwise the law of conservation of energy would be violated). To put it the other way round: A perpetual motion machine of the second kind would be technically feasible if one could get back to the starting point without delivering heat to the ambient. *The impossibility of a perpetual motion machine of the second kind operating with air as a work medium is hence not rooted in the impossibility of converting ambient heat to mechanical work at a ratio of 1:1, but in the incapability of compressing the gas isothermally without creating heat.*[1]

A similar proposition holds true for electrostatics. Ambient heat can be converted to electric work when the electric field between the plates of a capacitor is disappearing, resulting in an adiabatic cooling of the dielectric material between the plates (electrocaloric effect). After the dielectric has regained its initial temperature (that of the ambient) by means of a heat flow from the ambient, ambient heat has been completely converted to electric work (performed -during the discharge- in the electric circuit the capacitor is part of) as a net result.[2]

As with the gas in the cylinder, a perpetual motion machine of the second could be feasible if one managed to return to the starting-point without creating heat. Simply re-charging the capacitor would, however, not be an appropriate method, as the dielectric is warmed up by the same (now reversed) electrocaloric effect.

In other words: *The task to be solved consists in increasing the potential of a capacitor (that comprises a dielectric) without generating heat in the dielectric as an unwelcome side-effect.*

An attempt worth looking at might be the following: Consider a parallel-plate capacitor filled with vacuum (in order to avoid edge effects, that parallel-plate-capacitor shall be composed of two concentric spheres the diameters of which differ only slightly from each other). The voltage across shall be V_0. The attractive force experienced by each plate shall be called F_0. Let us now replace the vacuum by a dielectric with a relative permittivity K of well above unity. The voltage across the capacitor shall be kept invariant (at V_0). Will there be a change in the attractive force (now called F_1)? According to any textbook, the force will be increased by the factor K (permittivity of the

---

[1] See a standard textbook like R.W. Pohl: Mechanik, Akustik und Wärmelehre, 18th edition, Berlin 1983, p. 336 (my own translation):*"The isothermally working air compressor engine is a machine that converts the thermal energy received from the ambient; in an ideal situation the efficiency is 100%. The conversion occurs while the compressed air is relieved from its pressure. Now we add as a new remark: The relief from pressure increases the entropy of the air. ... This increase in entropy is a permanent and relevant change that the work medium is subject to when isothermally yielding work."*

[2] One might tend to raise an objection to that description by arguing that all the electric work is the result of the conversion of *energy stored in the electric field* (between the two plates) rather than the result of the conversion of *heat*. However, at least in case we are using a common model dielectric in which the polarization charges are are substituted by regular charges on the surface (with no charges in the interior), that criticism cannot be convincing: With a given voltage across the capacitor, the field between the plates has the same structure and strength as if the dielectric were vacuum, and does hence contain the same amount of energy. Yet the electric work performed by the discharging capacitor can be many times greater than that of the discharging capacitor filled with vacuum. To account for this excess in electric work, the role of ambient heat is indispensable.

dielectric). As Feynman [3] has pointed out, this holds true only in case the dielectric is a liquid. If it is a solid material, the attractive force is counteracted by a virtual pressure exerted on the plate by the solid dielectric in very much the same way as the weight of our bodies is counteracted by the virtual pressure of the ground against our feet. But even with a liquid dielectric (in which the capacitor is immersed), the result is somewhat puzzling: Though a **K** times greater charge sits on each plate, that (free) charge is partly neutralized by polarization charges appearing on the surface of the liquid (in a common model dielectric there are no other charges than these in the whole dielectric). So, when increasing the mutual distance between the two plates, the same *effective* charge is moved as would be moved in case we had not replaced the vacuum by the liquid dielectric. With the effective charge being equal in amount to that in the vacuum-filled capacitor, one would (intuitively) expect F_1 to be equal to F_0. So, how is the increase in force accounted for?

Fig.1

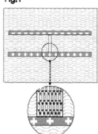

An explanation is revealed in fig. 1. In order to avoid errors by not giving consideration to the effect of the dielectric on itself, a tiny section of both the lower plate of the capacitor and the liquid is cut out and sealed off by vertical walls. The electrostatic force exerted on the free charge that sits on that tiny section of the lower plate is **K** times greater than it would be in case the dielectric were vacuum (at a given voltage V_0). This is due to the fact that the external field (that is the field generated by all charges outside of that section) acting on the depicted free charge is the same as it would be in case the dielectric were vacuum, with the depicted free charge itself, however, being **K** times larger in amount.

This result is not modified by the presence of the dielectric inside the section. The net force exerted on the column of dielectric (enclosed in the cut-out volume) by the external field is zero: As the external field inside that volume is homogeneous, all the dipoles (the dielectric is made up of) experience a torque only, and no translational force. As a consequence, the contribution (made by the column of dielectric) to the total force exerted on the cut-out section of the capitor plate is zero (we neglect the hydrostatic force generated by *gravity*).

Thus, when increasing the potential energy of the charges by doubling the mutual distance between the plates (with the charging wires disconnected), **K** times more mechanical work has to be spent (compared to a capacitor filled with vacuum). Since the *increase* in potential energy (= capacity of yielding electric work when eventually discharging the capacitor) brought about by that widening of space between the plates is also augmented by the factor **K** (compared to a capacitor filled with vacuum) due to a **K** times greater amount of free charge sitting on the plates, the quantity of electric work yielded (during the eventual discharge of the plates) matches the sum of electric and the mechanical work invested. This is why the work cycle does *not* represent a perpetual motion machine of the second kind, and the total account of heat added to and taken away from the dielectric must break even. As the electrocaloric effect of adiabatic *cooling* (during the discharge), due to the doubled volume of the dielectric between the plates, was twice as great in amount as had been the electrocaloric effect of adiabatic *heating* (during the charging of the capacitor), nothing else but the generation of heat during the process of moving the plates away from each other can be responsible for the balanced heat account.

Fig.2

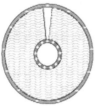

Things turn different if we consider fig. 2, which shows a spherical capacitor consisting of two concentric spheres of extremely unequal diameter. The

[3] Lectures on Physics II, 10-5

outer sphere is grounded. The voltage across that capacitor shall again be V_0. We will start with vacuum as a dielectric. Analogous to the parallel-plate capacitor, a tiny sector of the inner sphere's surface (and of the adjoining liquid dielectric) is cut out and sealed off by radial walls reaching to the outer sphere. The radial and outward (electrostatic) force experienced by the tiny section of the inner sphere shall again be called F_0.

We are interested in finding out whether or not this force will be affected by filling the -previously empty- interior of the spherical capacitor with a liquid dielectric. (The inner sphere shall be kept connected to the battery, so that the voltage V_0 stays invariant.) After the capacitor has been completely filled, the free charge in the tiny sector (on the surface of the inner sphere that is cut out by the radial walls) is K times greater in amount than it was prior to the insertion of the liquid dielectric, while the external field is invariant. Does this result in an K times greater outward force (which we call F_1) experienced by the tiny part of the surface of the inner sphere (that is cut out by the radial sector walls)? The answer is in the negative. The external field acting on the dielectric (within the sealed-off volume) is inhomogeneous. This is why every dipole is experiencing a radial force toward the center of the spherical capacitor. The individual forces sum up to build an inward pressure against the tiny part of the surface of the inner sphere, thereby counteracting the outward radial force experienced by the free charge. Thus, when reducing the diameter of the inner sphere to ½ (after having disconnected the charging wire), the mechanical work spent on the shrinking process is *less* than K times greater than it would be in case the capacitor were filled with vacuum. The *increase* in potential energy (= capacity of yielding electric work when eventually discharging the capacitor) brought about by that reduction of diameter is, however, augmented by the factor K (compared to the case in which the capacitor undergoes the same procedure while being filled with *vacuum*). Hence the quantity of electric work yielded in the cycle (consisting of charging the capacitor as a first step, making the inner sphere shrink -with the charging wire disconnected- as a second step, discharging the capacitor as a third step, and making the inner sphere expand to its original size as a fourth step) does no longer match the sum of electric and the mechanical work invested, but is in excess of that sum. This excess has to be at the expense of heat (otherwise the law of conservation of energy would be violated). That is to say: When charging the capacitor and reducing its diameter thereafter, the total amount of heat produced (and transferred to the ambient) must be smaller than the amount of heat converted to electric work during the eventual discharge. The electrocaloric effects are no longer symmetrical, and we are facing a Second Law violation.

This has been verified experimentally. By an arrangement described further below, the outward pressure of the inner sphere of a spherical capacitor (at a given voltage across the capacitor) was found to be the same no matter if the capacitor was filled with a liquid dielectric or with air, provided the diameter of the outer sphere is great enough. Thus it has been proved that physics textbooks are incorrect in asserting that all forces between charged conductors, at a given voltage, are increased by the factor K if these conductors are immersed in a liquid dielectric, independent of their arrangement.

It should be noted that for a Second Law violation to be avoided, the radial (inward) force experienced by the liquid dielectric in the sealed-off volume (as an effect of the external field) would have to be *zero* - which, however, would be incompatible with the undisputed fact that every single dipole is subject to a radial, inward force.

2. Quantitive description of the Second Law violation

A. Preliminary considerations: Quantitative description of the Electrocaloric Effect

a) If, for a short period of time, a single dipole is subject to a homogeneous electric field, its kinetic

energy can either be increased or reduced: It is *increased* in case ist random rotary motion (which, in turn, is due to the distribution of the thermal motion among more than three degrees of freedom) is performed in accordance with the torque created by the field. It is *reduced* in case its rotary motion is performed against the torque. The increase or decrease in kinetic energy shall be called dW_{dipol}. The term q_{de} denotes the charge at each end of the dipole. The scalar p_E denotes that component of the dipole moment $\vec{p}$ (vector) which is parallel or anti-parallel to $\vec{E}$ (vector), with the dipole moment $\vec{p}$ (vector) being defined as the product of q_{de} and the distance $\vec{r}$ (vector) between the two ends of the dipole (pointing from the negative charge to the positive charge). Each end of the dipole undergoes a displacement, but only that component of the displacement which is parallel or anti-parallel with the field $\vec{E}$ is what matters when the work shall be determined. That displacement is denoted as $1/2\ r_E$. F (taken as a scalar) denotes the force acting on each end of the dipole. It is always parallel or anti-parallel to the field $\vec{E}$. Then we have:
(1)

$$dW_{dipole} = 2\ F\ d(\frac{1}{2}\ r_E)$$

$$= \frac{F}{|q_{dt}|}\ |q_{de}|\ dr_E = \frac{F}{|q_{dt}|}\ dp_E = E\ dp_E$$

The term dW_{dipole} is defined as positive if the dipole is rotating *with* the field; it is defined as negative if the dipole is rotating *against* the field. The term dp_E might be different for every single dipole. The field E within the sealed-off section in fig. 1 can be separated into an external field E_{ext}, and a local field E_{local} the sources of which are located within the sealed-off section. The internal work within the dielectric (in the sealed-off section of fig. 1) would then be:
(1a)

$$dW_{intern} = \Sigma dW_{dipole} = E_{ext}\ \Sigma dp_{E_{ext}} + \Sigma E_{local}\ dp_{E_{local}}$$

The last sum (which can be defined as a function of E_{ext}) can be neglected, as it is zero for any value of E_{ext}.

The sum of all dipole moments $\vec{p}$ (vector) within a unit volume is called $\vec{P}$ (vector). That vector is always parallel and proportional to the external electric field.. For $\vec{P}$ we thus have (as an empirical law):
(2)

$$\vec{P} = (K - 1)\epsilon_0 \vec{E}$$

The term ϵ_0 denotes the dielectric constant. Thus (1a) can be converted to
(2a)

$$dW_{intern} = \Sigma dW_{dipole} = E_{ext}\ dP_{E_{ext}}$$

which (by making use of Equation 2) turns into:
(3)

$$dW_{intern} = E_{ext} \, dP_{E_{ext}} = \vec{E}_{ext} \cdot d\vec{P}$$

$$= \vec{E}_{ext} \cdot d[(K-1)\epsilon_0\vec{E}_{ext}] = (K-1)\epsilon_0 E_{ext} \, dE_{ext}$$

giving the internal work per unit volume [4]. Thus the total amount of internal work due to a change in the external field is given by the following integration:
(4)

$$W_{intern} = \int_0^E (K-1)\epsilon_0 E \, dE + \sum E_{local} \, dp_{E_{local}} = \frac{(K-1)\epsilon_0}{2} E^2$$

The integral is formed from E to 0 (rather than from 0 to E) if the field is fading. Then Equation 4 gives the amount of kinetic energy of the dipoles that disappeared as a result of the disappearance of the field.

If **K** is great enough, almost all of the energy provided by the battery (that energy being $1/2 \, K \, \epsilon_0 E^2$) is converted to heat. By the same token, almost all of the electric work done when a capacitor (filled with a dielectric) is discharged is at the expense of heat!

b) A cross-check shall confirm this result. When a capacitor (filled with a dielectric) is being charged, the bound charges within the dielectric undergo slight displacements that sum up to form a real current I_{real} (t), the reality of which becomes obvious as it makes a magnetic field as does as wire current (thus superposing and increasing the magnetic field brought about by the change in the electric field between the capacitor plates). With that current being a real current, it obeys Ohm's law. In order to apply Ohms' law, the dielectric is considered to be a resistor. For that resistor, we get:

---

[4] The same result is arrived at by Georg Joos: Theoretical Physics, Dover Publ., New York, 3rd edition, 1986, chapter XIII, par. 2, p.289/290, who also, when describing the energy density **u** in space filled with a dielectric, distinguishes between the change in energy of the vacuum and the change in the energy of a material body (dielectric). Starting from:

u = 1/2 K ϵ_0 E^2

he proceeds by developing **du/dE**. The result is multiplied by **dE**. He gets

du = K ϵ_0 E dE = (χ+1) ϵ_0 E dE
= ϵ_0 E dE + P dE = ϵ_0 E dE + E dP

with χ being equal to **K-1**, and with dipole moment density **P** being equal to **E** times χ ϵ_0. The last term **EdP** on the very right side (which matches with the expression of the change in internal energy found in our Equation 3) is labeled as „change in the energy of the material body", while the first term is labeled as „change in the energy of the vacuum".

I

(4a)

$$\frac{dW_{electr}}{dt} = I_{pol}(t)\, V(t)$$

which can be turned into [5]

(4b)

$$dW_{electr} = \frac{dq_{pol}}{dt}\, V(t)\, dt = dq_{pol}\, V(t)$$

Since the voltage V across the capacitor is proportional to the amount of polarization charge q_{pol} that has passed through the sectional area of the dielectric, V can be substituted by the product of q_{pol} times a constant $\bar{C}_1$. Thus (4b) turns into

(4c)

$$W_{electr} = \int_1^2 dW_{electr} = \int_0^{q_{pol}} C_1 q_{pol} dq_{pol} = \frac{1}{2} C_1 q_{pol}^2 = \frac{1}{2} V q_{pol} = \frac{1}{2} Eh\, q_{pol}$$

The term h denotes the distance between the two capacitor plates. Multiplying and dividing by the surface area A of each plate results in
(4d)

$$W_{electr} = \frac{1}{2}\frac{q_{pol}}{A}\, E\, hA = \frac{K-1}{2}\frac{q_{eff}}{A}\, E\, hA = \frac{(K-1)\epsilon_0}{2}\, E^2\, hA$$

The term q_{eff} denotes the effective charge (formed by the free charge plus the polarization charge). Dividing the last term of (4d) by the volume hA of the dielectric gives the electric work per unit volume just as in (4).

B. Absence of heat generation in a liquid dielectric (while increasing the potential of a conductor) as a consequence of Maxwell's Laws

As pointed out above, a Second-Law-violation would materialize if one managed to increase the potential of a charged conductor (immersed in a liquid dielectric) without generating heat in the dielectric as an unwelcome side-effect.

When making a spherical conductor shrink that is immersed in a "sea" of liquid dielectric, no

---

[5] In order to apply Ohm's law, it is, in addition, required that no form of energy other than heat is generated by the current. This is, of course, not true for the polarization current, which is building up its own magnetic field. But the energy stored in that field is turned into heat when the polarization current, at the maximum state of polarization, is dying out. This is why the error is being corrected „automaticly".

o

magnetic field **B** can emerge -simply for reasons of symmetry-, though the electric field is changing and charge is moving. Any magnetic field would define a privileged direction, for which there is no justification in a radial-symmetrical "world". The two mathematical "sources" of a magnetic field (see the right side of Maxwell's fourth equation as displayed below) just cancel.

Considering a surface section **A** (cut out by the closed path s) of a thought sphere concentric with the spherical conductor, we have:
(5)

$$\oint_s \vec{B} \cdot \vec{ds} = 0$$

Then, according to Maxwell's fourth law, which is
(6)

$$\nabla \times \vec{B} = \frac{\vec{j}}{\epsilon_0 c^2} + \frac{\delta \vec{E}}{c^2 \, \delta t}$$

(E denotes the electric field or the electric flux per unit area, **j** denotes the current density, that is charge per second and per unit area, c denotes the speed of light), we get:
(7)

$$\left| \frac{I}{\epsilon_0} \right| = \left| \frac{\delta(\int_A \vec{E} \, d\vec{A})}{\delta t} \right|$$

The term **I** denotes the current through that surface section **A**. Replacing **I** by **dq/dt**, and the surface integral of **E** by **EA**, (7) turns into:
(8)

$$\left| \frac{\delta q}{\epsilon_0} \right| = \left| A \, \delta E \right|$$

An integration of (8) leads to
(9)

$$\left| \int_1^2 \frac{dq}{\epsilon_0} \right| = \left| A \int_1^2 dE \right|$$

$$\left| \frac{q}{\epsilon_0} \right| = \left| A \, \Delta E \right|$$

The charge **q** denotes the amount of charge that passed through **A** during the increase or decrease of the field.

Several of those thought surface sections **A** shall be „installed" in the cut-out volume filled with liquid dielectric (as shown in fig. 2) at different distances from the center of the sphere. Let us see why: The (macroscopic) external work yielded by a translational move of all the dipole-ends (**dq**$_{di-}$ or **dq**$_{di+}$) inside the sealed-off quantity of the liquid dielectric (when the inner sphere of fig. 2 is shrinking) amounts to:
(10)

$$|W_{ext}| = |\int_1^2 \int_{q_{di-}} \vec{E}_{ext}\, dq_{di-}\, \vec{dr} - \int_1^2 \int_{q_{di+}} \vec{E}_{ext}\, dq_{di+}\, \vec{dr}\,|$$

E_{ext} denotes the field created by charges outside the cut-out volume in fig. 2, **q**$_{di-}$ and **q**$_{di+}$ denote the total amount of negative or positive dipole charge in the volume considered. As long as **A** is situated in the interior of the liquid dielectric (at an invariant distance from the center of the sphere), the flux through **A** stays invariant though the spherical conductor is shrinking, resulting in a zero flow of charge through **A** according to (9). Obviously, in the interior of the dielectric the move of negatively charged dipole-ends is neutralized by a move of positively charged dipole-ends [6]. A change in flux through **A** is only taking place when the *interfaces of the dielectric* pass through **A** during the shrinking of the conducting sphere. Thus (10) converts to:
(11)

$$|dW_{ext}| = |\vec{E}_{ext\ (interface\ 1)}\, q_{pol\ (interface\ 1)}\, \vec{dr} - \vec{E}_{ext\ (interface\ 2)}\, q_{pol\ (interface\ 2)}\, \vec{dr}\,|$$

Interface 1 is the inner interface (dielectric/conducting sphere), interface 2 is the outer interface of the dielectric at an indefinite distance. The term q_{pol} denotes the polarization charge on the interfaces of the dielectric. As the external field E_{ext} reduces to zero at interface 2 (with the polarization charges being equal in amount to those at interface 1), the second term of (11) reduces to zero, too.[7]

$E_{ext\ (interface\ 1)}$ equals 1/2 **E**, with **E** denoting the field just above the surface of the sphere (in a hypothetical slot within the dielectric parallel to the field lines). This can be proved by some simple reflections: It is well known that, at a given charge density, the field just above or below an evenly charged plane is the same no matter how large the plane is, and no matter what special height above the plane is considered. With the field between *two* charged planes of opposite signs (inside a

---

[6] One might pose the question as to whether or not volume polarization charges exist in the interior of the (isotropic) dielectric. The answer is in the negative. The shortest proof of the absence of volume polarization charges is found in the fact that the increase in capacitance of a condenser filled with some dielectric depends on the material only, and not on the shape of the condenser. Therefore no field lines can originate or terminate in the interior of the dielectric (this is pointed out by Leonard Eyges: The Classical Electromagnetic Field, New York -Dover Publ.- 1980, p. 104).

[7] Equation 11 matches with the result found by L. Eyges (The Classical Electromagnetic Field, New York - Dover Publ.- 1980, p. 163): „This is the desired result: the force on the body is *as if* the external field acted on the polarization charges." See also another derivation of this result in A. Trupp: A Theoretical Paradox Related to Solid Dielectrics that Cannot be Resolved even after a Computer Simulation Using Finite Element Analysis, in: Proceedings ESA (Electrostatic Society of America) Annual Meeting 1999, p. 100-102.

parallel-plate-capacitor) being equal to the full charge density (divided by ϵ_0), the field above or below one single plane is equal to 1/2 of the charge density (divided by ϵ_0).

When considering a location in the interior of a charged sphere just below an arbitrarily chosen point on its surface (at a differentially small distance from the surface), the field at this location is zero (as it is in the interior of *any* conductor). But the field generated by the differentially tiny and therefore flat surface zone around the depicted point equals 1/2 of the surface charge density divided by ϵ_0. Obviously, this field is superposed and exactly neutralized by the field created by the vast rest of the sphere.

When considering a location just *above* that arbitrarily chosen point on the surface of the charged sphere (at a differentially low altitude), the field equals the full charge density divided by ϵ_0 according to the properties of the field geometry. It is as if all charge were concentrated in the center; so the field -which can be expressed as Coulomb force per unit charge- is
(12)

$$\frac{1}{4\pi\epsilon_0} \frac{q_0\, q}{r^2} \frac{1}{q_0} = \frac{q}{\epsilon_0 A_{sphere}} = E$$

With the flat surface zone (of differential size) around the depicted point being responsible for half of the magnitude of **E**, it is obvious that the field created by the vast rest of the sphere ($E_{restsphere}$) is again equal to 1/2 of the surface charge density (q/A_{sphere}) divided by ϵ_0, that is $E_{restsphere} = E_{ext\,(interface\,1)} = 1/2\,E$. Moreover, $dE_{restsphere}/dr$ must be zero at the surface.

It is only the field created by the vast rest of the sphere that counts (for determining the radial force acting on the cut-out volume of the sphere in fig. 2), since the field created by the (differentially) small amount of free charge on the cut-out part of the inner sphere (within the cut-out volume) is not able to set the contents of that volume in motion.

Not only the polarization charge, but also the free charge on the cut-out part of the inner sphere is subject to **1/2 E**. So the total work invested when making the inner sphere shrink amounts to
(12a)

$$dW_{mech} = \frac{E(q_{pol}+q_{free})}{2}dr = \frac{E\,q_{eff}}{2}dr = \frac{E}{2}\frac{q_{eff}}{A_{sphere}} A_{sphere}\; dr = \frac{\epsilon_0}{2} E^2 A_{sphere} dr = \frac{\epsilon_0}{2}E^2\, dV$$

which is equal to the work required in case the dielectric were vacuum or air (V denotes volume, not voltage).

Now it can be realized that no creation of heat can occur within the dielectric: The *net electrocaloric effect* (as regards the charging and the decharging of the spherical capacitor) amounts to (see Equation 4):
(13)

$$\Delta W_{intern} = \int_{\Delta V} \frac{(K-1)\epsilon_0}{2} E^2\, dV$$

with ΔV denoting the space „abandoned" by the sphere as a result of its skrinking (**V** denotes volume). The *net electric work* yielded in the cycle amounts to:
(14)

$$\Delta W_{elect} = \int_{\Delta V} \frac{K\epsilon_0}{2} E^2 \, dV \qquad 11$$

The *mechanical work* of making the inner sphere shrink amounts to:
(15)

$$W_{mech} = - \int_{\Delta V} \frac{\epsilon_0}{2} E^2 dV$$

The net gain in (technically usable) energy amounts to
(16)

$$\Delta W_{elect} + W_{mech} = \int_{\Delta V} \frac{(K-1)\epsilon_0}{2} E^2 \, dV$$

That net gain is equal to the net electrocaloric effect. In case a creation of heat took place during the shrinking process, the law of conservation of energy would be violated.

3. Description of the Experiment Conducted

A. The experiment was conducted by Ben Wiens Energy Science Inc., Coquitlam BC, Canada, on contract. The following equipment was used:

(1) Outer sphere 30.5 cm OD, aluminum, both halves grounded to electrical outlet ground wire
(2) Inner sphere 5.5 cm OD, brass
(3) Inner upper half-sphere support hardware, plastic
(4) High voltage power supply, Bertan, 15,000 volts at 3 volts monitor
(5) Scale, 4 beam balance 1/100 gram accuracy
(6) Dielectric fluid, DC-100, 99.9% paraffin
(7) 80 liter bucket to catch dielectric fluid spill
(8) Plastic outer sphere support stand, in bucket
(9) Dielectric fluid hand pump, plastic construction
(10) Inner lower half-sphere connected permanently to high voltage with thin copper wire
(11) Inner upper half-sphere connected permanently to high voltage with thick copper wire
(12) Outer sphere grounding wire connected to electrical plug ground

Tests:
(1) Monitor voltage=2.72
(2) Actual voltage=13,600 Volts

(3) Net force change in air=0.30 „grams"=0.0030 Newton
(4) Net force change in DC100 dielectric fluid=0.50 „grams" (=0.0050 Newton) or more to start, down to 0.30 „grams" (=0.0030 Newton) after 20 minutes.

The inner sphere of the spherical capacitor was divided into a lower and an upper half, with the upper half firmly connected to the outer sphere by plastic struts, and with the lower half suspended by a string that was attached to scales above the spherical capacitor. A hole at the „North Pole" of the (grounded) outer sphere allowed the string to penetrate the wall of that sphere. Both halves of the inner sphere were permanently in contact with a high voltage power supply of 13,600 Volts. In air, the electrostatic repulsive force which the lower half was subject to was 0.30 „grams", that is 0.003 Newton. The force in air with the charging wire removed was basically the same as with charging wire connected permanently, so this method produced just as reliable results. In the dielectric liquid, the electrostatic repulsive force (which the lower half was subject to) was 0.30 „grams", that is 0.003 Newton (just as in air), after a state of equilibrium had been attained.

In a second experiment, the attractive force between two parallel plates charged with electricity of different signs was measured in air and in paraffin oil (at an identical voltage).

1. Experimental Apparatus
(a) 80 liter plastic bucket
(b) 1/100 gram accuracy 4 beam balance
(c) Stainless Steel disks with sharply upturned edge about 210 mm diameter
(d) Rod to connect upper Stainless Steel disk to scale was brass threaded rod and wire hook to scale
(e) Grounding to upper Stainless Steel disk via grounding of all metal balance and linkages to upper disk.
(f) High voltage supply to lower Stainless Steel disk via insulated wire to Bertan high voltage supply
(g) Monitoring voltage was 2.72 volts=13,600 volts (3 volts=15,000 volts)
(h) Distance of flat portion of disks apart in tests in air and paraffin oil was identical and about 40 mm.
(i) Paraffin oil was DC100 and covered the upper disk completely in the test
(j) Upturned edges of disks were facing away from each other

2. Test in air
Attractive force=4.70 „grams" and steady over many minutes.

3. Test in paraffin oil
Attractive force=10.00 „grams" at 1 minute, 12.00 „grams" at 10 minutes, 12.80 „grams" at 120 minutes. Hence the equilibrium force was about 2.7 times greater in paraffin oil (as a liquid dielectric) than in air. This matches with the relative permittivity of paraffin oil, which ranges from 2.2 to 4.7 .

B. As with most liquid dielectrics, conductivity is much higher in paraffin oil than in air. Thus paraffin oil is far from being a perfect insulator, and can be considered to be a resistor. Does this imply an accumulation of free charge in the dielectric and hence an error in measuring the force of repulsion? The answer is in the negative. According to Ohm's law we have
(17)

$$ I = \frac{\Delta V}{R} = \frac{\Delta V\, A}{R_{specif}\, L} = \frac{\Delta V\, A}{L}\, \sigma $$

I denotes the current of free charge, ΔV the voltage drop along a bundle of stream lines within a resistor, L the length of the bundle of stream lines, A the average cross-sectional area of that bundle, R the resistance, R_{specif} the specific resistance, σ the specific conductivity (which is the reciprocal of the specific resistance). Since Ohm's law is still valid when all terms have differential dimensions only, (17) can be expressed as:
(17a)

$$ \frac{dI}{dA} = \frac{dV}{dL}\, \sigma $$

With the current density j being defined as dI/dA, and with the field E being dV/dL, we get
(18)

$$ \vec{j} = \sigma \vec{E} $$

The current density is proportional to the field inside the resistor at any spot considered.

As the divergence of $\vec{j}$ is zero *as soon as the current is steady* (the same amount of charge that enters a volume element per temporal unit must leave that element per temporal unit), the divergence of $\vec{E}$ must, too, be zero. Hence, as a consequence of Gauss' law, there can be no accumulation of charge in the interior of the resistor. This has been labeled as a „fundamental law of electricity" in standard textbooks. [8]

Things are different only if σ varies along the electric circuit. At the interface of two different resistors, **j** has to stay invariant, while σ is undergoing a change. This implies a reciproke change in **E** according to (18). Thus, when imagining a right-angled volume element that comprises both resistors so that the current enters through the left side and leaves through the right side, the divergence of **j** is still zero, but the divergence of **E** is different from zero. Therefore, according to Gauss' law, some accumulated charge must sit in that volume element.

That charge is the same in amount as if the dielectric were vacuum: With the high specific conductivity of the brass (the inner sphere is made of) as the conducting material on the one hand, and with the weak current density provided for by the liquid paraffin (as a fairly good resistor material) on the other hand, the field **E** inside the brass is practically zero according to Equation 18. Therefore the flux of **E** through the inner side of a (thought) right-angled volume element that comprises both the brass and the dielectric is zero. Since the radial geometry and the strength of the field between the two spheres of the capacitor are the same as if the dielectric were vacuum (due to the absence of free charges in the interior of the dielectric), the flux through the *outer* side of the volume element is also the same as if the dielectric were vacuum. Then, according to Gauss' law, the effective charge within the volume element, too, has to be the same in amount as if the dielectric were vacuum.

The absence of free charges in the interior of resistors, and the fact that the distribution of effective charge on the surface of a metal body (when permanently connected to a pole of a battery and immersed in a liquid of high resistance) is the same as if the metal body were embedded in vacuum, entails a spatial distribution of the potential (in an entirely resistor-filled space surrounding the conductor), which, at a given potential of the conductor, is identical to the spatial distribution (of the potential) generated if the space were filled with vacuum. This property is being used in experimental physics to determine the spatial distribution of the electrostatic field generated by irregularly shaped bodies in space [9]: A model of the body made of metal is immersed in a bath containing an electrolytic liquid. The metal body is connected to one pole of a power supply, and the far away (metal) container that holds the liquid is connected to the second pole of the power supply. A probe with an insulated wire is introduced into the liquid, with the other end of the wire connected to the container. The voltage given by a dynamic voltmeter is an expression of the potential of the probed

---

[8] See G. Bruhat: Cours d'Electricite, 3rd edition, Paris 1934, p. 223 (my own translation): „ *We have already stated that the vector i satisfies the condition div i=0 under a permanent regime. It follows from Ohm's law that one also has div E=0, and since, according to Poisson's law, one has div E=...=4pi q, it follows as a fundamental law that the electric volume density, when a system of conductors has arrived at a permanent regime, is zero at all points of the conductors; the electrification of the conductors is purely superficial"*. See also Georg Joos: Theoretical Physics, Dover Publ., New York, 3rd edition 1986, chapter XIV, par. 1, p. 294/295.

[9] See Georg Joos, op. cit.; Grimsehl: Lehrbuch der Physik, Vol. 2, 13th edition, Leipzig 1954, p. 170; standard experimental devices for two-dimensional electrostatic problems that use thin sheets of solid resistor material in connection with probes and voltmeters are being sold by companies that supply lab equipment to schools, colleges and universities, see for instance the catalogue of „PHYWE", based in Goettingen, Germany.

location.

C. It can be realized that it is the *state of a steady currents* -and not the initial state of time varying currents- which counts when determining the electrostatic force of repulsion. Only the state of steady currents is a state that guarantees the absence of free charges in the interior of the resistor (dielectric). The fact that an equilibrium (steady current through the dielectric) is arrived at only after some time has elapsed can be accounted for as follows: When undergoing a fast increase in voltage by a power supply, capacitors can accumulate some excess charge (= more than in a state of equilibrium) for a short time -during which the surface of the conducting material (plates and wires) does *not* form an equipotential surface- in very much the same way as the surface of a liquid may temporarily rise above normal (=equilibrium) height by the action of a wave. This can be illustrated by considering two vacuum-filled capacitors (of identical capacitance) in a serial arrangement, with only one of the two being charged. As soon as the gap in the connecting wire between the two capacitors is being closed by a switch, charge is flowing from the first to the second capacitor through an ideal wire with no resistance. If the flow came to a standstill as soon as the charge is evenly distributed among the two capacitors, the law of conservation of energy would be violated: Each of the two capacitors would contain only 1/4 of the energy previously stored in the first capacitor, since the energy of the field between the plates is proportional to the *square* of the charge density. In order to avoid a violation of the law of conservation of energy, the charge is oscillating from one capacitor to the other and vice versa, and the divergence of the current density $j=I/A$ is different from zero. As real wires *do* have a resistance different from zero, the oscillation is dampened until it fades out [10].

Back to the spherical capacitor filled with a dielectric: When there is a charge wave (making **div** $\vec{j}$ differ from zero) which travels along the circuit, that wave of accumulated charge is only partly reflected at the interface sphere/dielectric. Some charge (say, negative charge) enters the dielectric and is pushed forward by the existing field as long as the sphere is still charged with electricity of the same negative sign. As soon as the electricity on the sphere changes to positive, the charge that had entered the dielectric is drawn back by the (then existing) field. However, since the wave is dampened both by the resistance of the conductor and by the previous loss of negative charge to the dielectric, only a fraction of the negative charge inside the dielectric manages to leave the dielectric during the phase in which the charge on the sphere is positive. Thus some residual negative charge remains in the dielecric after a full wave (2 π) has been completed. That residual accumulation of charge will migrate along the steady field which is established as soon as the oscillation in the conductor has come to a standstill. Due to the poor conductivity of the dielectric, it may take several minutes or more for that cloud of accumulated charge to traverse the dielectric.[11] Not before the disappearance of that charge will a state of dynamic equilibrium be attained.

d) The permittivity of paraffin oil **K** ranges from 2.2 to 4.7. Hence, in order to avoid a violation of the Second Law, the repulsive force would have to be at least 2.2 times 0.0030 = 0.0066 Newton rather than 0.0030 Newton (which is the force in air). Only then would the result match with the textbook version of Coulomb's law in dielectrics, which postulates an increase in forces between

---

[10] See A.D. Moore: Electrostatics, 2nd edition, Morgan Hill 1997, p.106: *"When a capacitor discharges by way of a spark the current in the spark is typically oscillatory. It surges back and forth quite a few times at high frequency before it dies out."*

[11] Conductivity in highly insulating liquids is caused mainly by „foreign" ions that are present due to the impurity of the liquid; in order to remove *those* ions, electric fields have to exert their influence on the liquids over a period of several days. See: R.W. Pohl: Elektrizitätslehre, 18th edition, Berlin 1983, p. 177. So it is no surprise to find an equilibrium only after several minutes or even hours.

conductors by the factor **K** (=the relative permittivity of the dielectric) as a result of the presence of the liquid dielectric.

The drag created by the stationary current of free electrons or negative ions that "bump" against the molecules of the dielectric does not give rise to a modification of the expected result. It is just to the contrary: Without those "bumps", the positively charged ions in the dielectric would transmit a net force to the liquid no matter if they are immobile or being dragged through the liquid themselves. It is by the effect of those "bumps" that this force on the liquid is neutralized.

A slight modification of the expected result is, however, brought about by the fact that the diameter of the outer sphere is not infinite, but only 5.5 times greater than the diameter of the inner sphere. Hence the expected force is about 4% greater than in case of air as a dielectric, that is 0.0031 Newton.

In order to measure forces by the scales used, the object (lower half of the inner sphere, upper plate of the parallel plate capacitor) had to undergo a (slight) vertical displacement within the liquid. Dynamic viscosity of the liquid was nothing to worry about, since internal friction is proportional to the velocity of the object, so this effect can be evaded by simply reducing the speed of motion through the liquid. But sheer forces, which are zero in a liquid when the relative speed of the object and the liquid is zero, can (in principle) be different from zero -even if everything is at rest- due to electric forces between the dipoles of the dielectric (resulting in the phenomenon of pressure being dependent on direction). So it was not certain *a priori* whether or not the object would resist a displacement unless a minimum force of considerable amount (in the order of the expected force of electrostatic repulsion) would be applied. This did not happen.

Review of

"Second Law violations in the wake of the Electrocaloric Effect in liquid dielectrics", by Andreas Trupp

*This paper is unacceptable for **Physical Review E**.*

The reasons for this assessment are as follows:

1) The vast majority of the paper is a narrative which is alternately pedagogical and speculative. Large parts of the narrative describe basic and/or textbook physics. Such substantial pedagogical treatments are inappropriate for journals devoted to research into new or previously unexplained phenomena. Other portions of the narrative are speculative discussions of possible physical phenomena which might be at play. This is also inappropriate for a research journal.

2) The author provides an unacceptably brief experimental section 3A. The level of this work is not suitable for Physical Review. The experimental scientists who completed the work should be identified. Dielectric materials of undetermined properties are utilized. Systematic measurements using the same apparatus for a variety of conditions (voltages, timing of variations, dielectrics) have not been made (at least not reported) to allow separation of geometrical errors or apparatus flaws to be distinguished from physical phenomena. There is inadequate experimental

ιʊ

information to support the speculative analysis.

N. B. A.
Editorial Board Member
Physical Review E

Any lab interested in repeating the experiments will be provided with the experimental apparatus free of any charges or costs. Please contact me under my private email address:

atrupp@aol.com

Phone: (011)49-3338-760937

Four Paradoxes Involving the Second Law of Thermodynamics

By D.P. Sheehan

Appendix G follows on pages 155-160. This paper first appeared in the *Journal of Scientific Exploration*, vol. 12, no. 2, pp. 303-314, 1998. Reprinted with permission.

Four Paradoxes Involving the Second Law of Thermodynamics

———— D.P. Sheehan* ————

Reprinted with permission from the *Journal of Scientific Exploration* (Vol. 12, No. 2, 1998).

Abstract

Recently four independent paradoxes have been proposed which appear to challenge the second law of thermodynamics.[1-12] These paradoxes are briefly reviewed. It is shown that each paradox results from a synergism of two broken symmetries—one geometric, one thermodynamic.

Author Note

This article was first published five years ago in the *Journal of Scientific Exploration* (*JSE*). Since then the field of second law challenges has undergone dramatic growth. Over the last ten years, over forty articles on the subject have been published in the general scientific literature, more than during any other decade of the second law's 150 year history [*e.g.*, see References 1-35]. The pace is accelerating. Several independent research groups worldwide are pursuing challenges and the movement from purely theoretical proposals to laboratory experiments has begun in earnest. The first international conference on the subject of second law limits was convened in San Diego in 2002, bringing together more than 120 researchers from twenty-five countries;[13] additional conferences are planned. The first book on the subject commissioned by a major scientific press is due to be published in the next year.[14] Related books are already in print.[15,16] Taken together, these developments signal a fundamental change in how the scientific community views the second law. A paradigm shift may be on the horizon.

Since its publication in *JSE*, the class of second law paradoxes to which this reprinted article speaks has expanded from four to five[17] distinct challenges and a broader understanding has developed, particularly with respect to the gravitational paradox and to the general physical mechanism upon which all are based. Although some details of this reprinted article are now dated, it gives an introduction to most of the challenges investigated at the University of San Diego over the past decade. Author Notes have been added to this text to shore up deficiencies; by no means are they exhaustive.

I would like to thank *JSE* both for its help in originally publishing this work and for its permission to reprint it. I also belatedly thank the *JSE* referees for their critical comments.

Introduction

The second law of thermodynamics is empirical. It has no fully satisfactory theoretical proof. This being the case, its absolute validity depends upon its continued experimental verification in all thermodynamic regimes. Physical processes involving broken symmetries have been the standard touchstones by which its validity has been tested. Recently, four simple paradoxes have been posed which appear to challenge it.[1-12,17] In each, the universe consists of an infinite isothermal heat bath in which is immersed a blackbody cavity. Within each cavity steady-state, non-equilibrium thermodynamic processes create spontaneous asymmetric momentum fluxes which are harnessed to do steady-state work. If one demands the first law of thermodynamics be satisfied by these systems, then apparent contradictions of the second law result. Laboratory experiments and numerical simulations have corroborated theoretical predictions and have failed to resolve the paradoxes in favor of the second law. In this paper, it is shown that a broken symmetry in each of these four systems' thermodynamic properties allows asymmetric momentum fluxes to arise spontaneously, and that these can be harnessed to perform work utilizing a second broken symmetry in each system's geometry. By illuminating this characteristic shared by these paradoxes, it is hoped that their resolution will be expedited.

It may be thought that asymmetries such as these are thermodynamically forbidden and that each system must relax to an equilibrium characterized by spatial homogeneity. This is not the case. In fact, "equilibrium" does not forbid spatial gradients so long as they are *steady-state* ones. For example, the asymmetric momentum fluxes to be introduced shortly in Systems II, III, and IV are no more than steady-state pressure gradients.

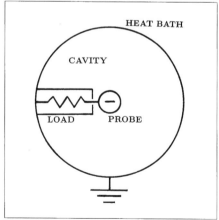

Figure 1. Schematic of paradoxical System I. The probe bias is assumed negative.

Equilibrium (steady-state) pressure gradients are ubiquitous in nature. For instance, they are standard features of gravitationally-bound, isothermal, static atmospheres on idealized planets. In a uniform gravitational field, one can write the gas pressure as a function of vertical height, z, as $p(z) = p_o exp[-mg(z-z_o)/kT]$, where m is the mass of the gas molecule, kT is the thermal energy, g is the local gravitational acceleration, and p_o is a fiduciary pressure. Clearly, this atmosphere possesses a vertical pressure gradient at equilibrium. Similarly, the pressure gradients in Systems II-IV are steady-state structures, but unlike the atmospheric gradient which is *static* and due to a static potential gradient (gravity), these pressure gradients are *dynamically* maintained by the continuous effluxes from two surfaces having different activities toward the cavity gas. Furthermore, these pressure gradients can do work.

In the next section, the four paradoxes are reviewed briefly and in Section 3 they are discussed in the context of broken symmetries. Detailed descriptions of the paradoxes are found elsewhere.

Four Paradoxes

System I (Plasma I)

System I[2-4] consists of a blackbody cavity containing a low-density plasma and an electrically conducting probe connected to the walls through a load, as shown in Figure 1. The load may be conservative (*e.g.*, a motor) or dissipative (*e.g.*, a resistor). The probe and load are small enough to represent minor perturbations to the cavity properties. The walls are grounded to the heat bath both thermally and electrically ($V_{ground} = 0$). The potential between the bulk plasma and the cavity walls—the plasma potential, V_p—may be positive or negative depending on the work function and temperature of the walls, and the plasma type and concentration. For an electron-rich plasma and in the absence of any net current to the plasma or walls, V_p may be estimated by equating the Richardson emission, J_R, from the walls to random electron flow from the plasma into the walls:

$$J_R = AT^2 exp\left(-\frac{e\Phi}{kT}\right) exp\left(\frac{eV_p}{kT}\right) = \frac{nev_e}{4} \qquad (1)$$

Here ϕ is the wall's work function, T is temperature, V_p is the plasma potential, v_e is the average electron thermal speed, k is the Boltzmann constant, m_e is the electron mass, n is the plasma particle density, and A is the Richardson constant (about $6 - 12 \times 10^5$ (A/m^{2}K^2) for pure metals). Under either equilibrium or non-equilibrium conditions, V_p will be non-zero except for very specific plasma parameters; in particular, $V_p = 0$ at the critical density, $n_c = (4AT^2/ev_e) exp[-e\phi \backslash kT]$, derived from Equation 1.

The probe will achieve a potential with respect to the plasma and walls depending on its temperature, resistance to ground (load resistance, R_L), and the current to it. Since it is nearly in thermal equilibrium with the walls, the probe is self-emissive and, therefore, electrically floats near the plasma potential so long as R_L is large. If $V_p \neq 0$, a current can flow continuously from the probe, through the load, to ground. This current represents an asymmetric momentum flux. The gener-

ated power may be expressed as $dW/dt = I_L^2 R_L \approx (V_p^2 / R_L)$. The entropy production rate is $dS/dt = (1/T)(dW/dt) \approx (V_p^2 / R_L T)$; this will be positive (negative) for a purely dissipative (conservative) load. Laboratory experiments corroborate this effect.[2] Note: this paradox is not limited to systems with thermionically emitting walls and probe; any plasma with a non-zero floating potential appears viable.[3]

If this system does steady-state work on the load while maintaining spatially steady-state temperature and species concentration profiles, and if the first law of thermodynamics is satisfied, then a paradox involving the second law naturally develops. Formally, the first law states: $[\Delta Q - \Delta W]_{hb} = -[\Delta Q - \Delta W]_c$, where hb refers to the heat bath and c refers to the cavity. The heat bath supplies heat, but does no work, so $\Delta W_{hb} = 0$. If the load is conservative and each part of the cavity is at a steady-state temperature, then $\Delta Q_c = 0$. (It is assumed, without further justification, that there are no net phase changes or chemical reactions in the cavity.) Returning to the first law, since $\Delta W_{hb} = 0$ and $\Delta Q_c = 0$, this leaves $\Delta Q_{hb} = \Delta W_c$. The cavity does positive work, so $\Delta W_c = \Delta Q_{hb} < 0$; in other words, the work performed by the load is drawn as heat from the heat bath, a reasonable result.

Now consider the second law. Entropy is an additive thermodynamic quantity so the entropy change for the universe can be written: $\Delta S_{universe} = \Delta S_{hb} + \Delta S_c$. Since $\Delta Q_c = 0$, one has for the cavity, $\Delta S_c = \Delta Q_c/T = 0$. (Equivalently, one may argue that entropy is a state function and the closed cavity is in a steady state—having no net phase changes, chemical reactions, temperature or volume changes, the number of microstates available to it is fixed—thus the entropy of the cavity is time invariant, and so $\Delta S_c = 0$). With $\Delta S_c = 0$, one is left with: $\Delta S_{universe} = \Delta S_{hb} = \Delta Q_{hb}/T_{hb} < 0$. This violates the second law of thermodynamics, namely that for any spontaneous thermodynamic process, $\Delta S_{universe} \geq 0$. If one replaces R_L with a dissipative load, the second law is violated still, since a forbidden, permanent temperature gradient has been established between the load and the cavity ($T_{load} > T_c$). Note that this system is not in thermal equilibrium; this process is irreversible. In order to use validly equilibrium thermodynamic relations, the work must be performed "slowly." This can be achieved to any degree of precision desired by adjusting the load resistance. Similar arguments establish the remaining three paradoxes. Note also, neither this system nor the other three utilize standard thermodynamic cycles or a low temperature heat reservoir.

System II (Plasma II)

Paradoxical System II[5] is a mechanical analog to System I. Too, it consists of a blackbody cavity surrounded by the heat bath. The cavity contains a low-density ionizable gas, B, and a frictionless, two-sided piston (see Figure 2). As before, Richardson emission greatly exceeds ion emission for all surfaces, giving an electron-rich plasma with a negative plasma potential. The majority of the piston is of identical composition as the walls (surface type 2, S2), however, on one piston face is a small patch having a different work function (surface type 1, S1). It is small in the sense that it is relatively unperturbing to global plasma properties. The work functions of S1 and S2 and the ionization potential of B are ordered as: $\phi_1 \geq I$. P. $> \phi_2$. Plasma production is straightforward: electrons are "boiled" out of the metal (Richardson

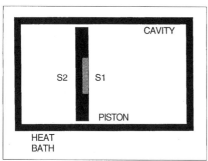

Figure 2. Schematic for paradoxical Systems II and III.

emission) and ions, created by surface ionization, are accelerated off the metal surface by the electron negative space charge. Ions, in turn, ease the electrons' space charge impediment, thus releasing a quasi-neutral plasma from the surface. Actually, if $V_p < 0$, this is essentially a charge-neutralized, low-energy ion beam leaving the surface. In fact, this plasma can be roughly considered to be an unmagnetized, three-dimensional Q-plasma with a sliding hot plate.[36,37]

The ordering $\phi_1 \geq I.$ $P. > \phi_2$ allows, with appropriate plasma density and temperature, and surface areas $((SA)_2 \gg (SA)_1)$, the following: surface 1 ionizes B well and recombines it poorly while surface 2 ionizes B poorly, but recombines B well. Surface 2 dominates plasma properties by virtue of its greater surface area $((SA)_2 \gg (SA)_1)$, therefore, the net flux of B to any surface is predominantly neutral B. Surface 1 will be relatively unperturbing to cavity plasma conditions if the $S2$ ion current into the plasma is much less than the total $S1$ ion current. The electron emission off $S2$ exceeds that off $S1$ by a factor $exp[(\phi_2 - \phi_1)/kT]$. The electron current density from each surface is given by Equation 1.

Because of the differences between neutral, electronic, and ionic masses and the different currents of each leaving $S1$ and $S2$, a steady-state asymmetric momentum flux density (a "net pressure difference," ΔP), is sustained between piston faces. It has been shown that this pressure difference is roughly

$$\Delta P = -\frac{p_{i,2} n_n kT}{2} - \frac{\pi}{4} \frac{m_e v_e}{e} J_{R,1} + m_i \frac{n_n v_n}{4} p_{i,2} \sqrt{-\frac{2eV_p}{m_i}} \quad (2)$$

where $p_{i,2}$ is the ionization probability of B on $S2$, n_n is the neutral density, $J_{R,1}$ is the Richardson current density from $S1$, m_i is the ion mass, and v_n is the neutral thermal velocity. The first, second, and third terms represent neutral, electronic, and ionic pressures, respectively. Laboratory experiments corroborate steady-state differential thermionic emission from different surfaces under blackbody conditions.[5] Numerical simulations, using realistic physical parameters, indicate the pressure effect is small, but significant.[2] If the piston moves slowly ($v_{piston} \ll v_n$) and performs work quasi-statically, it generates steady-state power, $(dW/dt)_{piston} = \Delta P (SA)_1 v_{piston}$, and

produces negative entropy at the rate, $dS/dt = (1/T)(dW/dt)_{piston}$. Notice that, even in the absence of a plasma potential, V_p, the paradoxical effect persists so long as the ionization probability of the two surfaces are distinct.

System III (Chemical)
Paradoxical System III[6-8] is the chemical-mechanical analog of System II. It consists of a blackbody cavity with piston into which is introduced a small quantity of dimeric gas, A_2. The cavity walls and piston are made from a single material, surface type 2 ($S2$), except for a small patch of a different material, surface type 1 ($S1$), on one piston face, as shown in Figure 2. (Note $S1$ and $S2$ here are distinct from those in System II.) The chemical model for this system assumes the following: a) the gas phase density is low such that gas phase collisions are rare compared with gas-surface collisions, however, it is sufficiently high that rms pressure fluctuations are small compared with the average pressure; b) all species contacting a surface stick and later leave in thermal equilibrium with the surface; c) the only relevant surface processes are adsorption, desorption, dissociation, and recombination; d) the fractional surface coverage is low, so adsorption and desorption are first order processes; e) A_2 and A are highly mobile on all surfaces and may be treated as a two-dimensional gas; and f) atomic and molecular species are retained sufficiently long on any surface to achieve close to chemical thermal equilibrium in the surface phase. These conditions are physically realistic and have been shown to be self-consistent.[6] For these conditions, it can be shown that, in principle, $S1$ and $S2$ can simultaneously desorb different ratios of A and A_2 in a steady-state fashion. However, since two A's together impart $\sqrt{2}$ times the impulse to the piston as does a single A_2 (all leaving in thermal equilibrium with the surface), asymmetric momentum fluxes can be sustained between the piston surfaces. (Another way to view this is: equipartition of energy does not imply equipartition of linear momentum.) The pressure imbalance on the piston faces can be used to perform work in a similar manner to System II.

For low surface coverage where desorption is a first order process, the desorption rate ratio for A and A_2, $R_{des}(A_2) / R_{des}(A) \equiv \alpha$, is given by:

$$\alpha \equiv \frac{R_{des}(A_2)}{R_{des}(A)} = \frac{n(A_2)}{n(A)} \frac{F(A)}{F(A_2)} exp\{\frac{\Delta E_{des}(A) - \Delta E_{des}(A_2)}{kT}\} \quad (3)$$

Here $\Delta E_{des}(A_j)$ is the desorption energy of A_j; $n(A_j)$ is the surface concentration of A_j (m^{-2}); and $F(A_j) \equiv (f / f^*)_{A_j}$ is a ratio of partition functions. f is the partition function for the species in equilibrium with the surface, and f^* is the species-surface partition function in its activated states. For real surface reactions, $F(A_j)$ typically ranges between roughly 10^{-3} - 10^4. Experimental values of desorption energy, ΔE_{des}, typically range from about 1 kJ/mol for weak physisorption up to about 400 kJ/mole for strong chemisorption.

The ratio α varies as $0 \leq \alpha \leq \infty$ depending on the values of the several variables in Equation 3. Experimental signatures of differential α's (some under quasi-blackbody conditions) are abundant.[38-41] If $\alpha_1 \neq \alpha_2$, and if the instantaneous fluxes of A and A_2 from $S2$ each greatly exceed those from $S1$ so that $S1$ can be treated as an impurity, i.e. $R_{des}(2,A_2)/R_{des}(1,A_2) \gg (SA)_1/(SA)_2$ and $R_{des}(2,A)/R_{des}(1,A) \gg (SA)_1/(SA)_2$, then a steady-state difference in momentum flux density (net pressure difference, ΔP) can be sustained between piston faces. Here $(SA)_j$ is the surface area of the jth surface. This pressure difference can be expressed:

$$\Delta P = \left(2 - \sqrt{2}\right) m_A v_A R_T(A) \left[\frac{\alpha_2 - \alpha_1}{(2\alpha_1 + 1)(2\alpha_2 + 1)} \right] \quad (4)$$

where $R_T(A)$ is the total flux density of A onto a surface, $R_T = [n(c,A)v_A + 2n(c,A_2)v_{A2}]/\sqrt{6\pi}$. Here $n(c,A_j)$ is the cavity concentration of A or A_2. In the limit that $\alpha_2 \gg 1 \gg \alpha_1$, the greatest pressure difference is obtained; it is roughly: $\Delta P \approx 0.3\ m_A v_A R_T(A)$. This pressure difference is steady-state since the dynamic chemical processes giving rise to it are steady-state. If this pressure difference is significantly greater than the statistical pressure fluctuations in the cavity, then, in principle, it can be exploited to do steady-state work. The power and entropy production rates here are the same as for System II. As for System II, the piston must move *slowly* compared with the thermal velocity of gaseous A_2. Note that, when the piston moves, the volume and surface phases for this system are not in equilibrium; in fact, they are in steady-state non-equilibrium.

This chemical system has been simulated numerically.[6] Closed-form, analytic rate equations have been developed and solved simultaneously using realistic physical parameters. Solutions confirm the possibility of this paradoxical effect; it is probably small—but significant—and appears viable over a wide range of physically accessible parameters. Laboratory systems displaying this effect are currently being sought. [Author Note: Laboratory experiments have been conducted that strongly corroborate the essential gas-surface reactions of this model.[8] Differential dissociation rates of molecular hydrogen were observed on high temperature tungsten and molybdenum surfaces under identical thermodynamic conditions (temperature, gas pressure).]

System IV (Gravitational)
[Author Note: The theory for this paradox has been superceded in several subsequent publications; the reader is directed to them.[9-12] Corroborative laboratory experiments for fundamental physical processes were carried out.[9]]

To introduce System IV, consider an everyday scenario: from the same height, drop a glass marble onto two different surfaces, for instance, a hard wood floor and a soft rug. The marble inelastically rebounds to different heights, demonstrating the different inelastic (endoergic) responses of the two surfaces. Inherently, these collisions are non-equilibrium processes. Analogous non-equilibrium behavior is observed on the atomic scale: it is well known that hyperthermal gas-surface collisions can excite energy states associated with internal degrees of freedom of either the collider or target—*e.g.* rotational, vibrational and electronic modes, phonons, plasmons—thereby rendering the collisions inelastic.[42-46] In fact, a number of standard surface diagnostics are based upon just such characteristic inelastic responses.[42-46] In contrast, at thermal equilibrium gas-surface collisions must, on average, be *elastic*, otherwise more direct contradictions with the second law arise. ("Hyperthermal" collisions are those with impact energies far above thermal energies—typically a few

tenths of an eV up to about 100 eV in energy.) Studies indicate energy transfer efficiencies from hyperthermal colliders to targets can range from a few percent to over 90% of incident atom kinetic energies.[42] Motivated by these observations, a simple, idealized system is considered: a strongly gravitating rod, whose ends have different inelastic responses to hyperthermal impacts by a particular gas, is placed at rest in a blackbody cavity with that gas. When steady-state is reached, gas continuously falls hyperthermally onto the rod, inelastically rebounds to different degrees from the rod ends, and is rethermalized in the blackbody cavity. The particle fluxes to and from both rod ends are identical, but the momentum fluxes are different, giving rise to a net force on the rod. If released, the rod accelerates in the direction of the net force and, in principle, can be harnessed to do mechanical work.

The idealized system consists of: (i) an infinite heat bath; (ii) a large, spherical blackbody cavity; (iii) a low density gas in the cavity; and (iv) a rod gravitator. The rod (length $2L_g$) has symmetric mass density $\rho(x) = \rho(-x)$ about its center at $x = 0$, but its end surfaces ($S1$ and $S2$) are composed of two materials distinct in their inelastic responses to gas atoms (mass m_A). In other words, for $S1$ and $S2$ one can write the inelastic response functions as distinct: $v_f(1,v_i) \neq v_f(2,v_i)$. The inelastic response function for surface j, $v_f(j,v_i)$, maps the velocity of a particle before impact, v_i, onto its velocity after impact, v_f.

The rod represents a minor perturbation to the overall cavity properties. Its gravitational scattering length L_s is much smaller than the cavity scale length, L_c. As a result, N_s, the ratio of the average number of wall collisions (N_{wall}) to the number of rod collisions (N_{rod}) it undergoes, is large; that is, $N_s \equiv N_{wall}/N_{rod} \approx (L_c/L_s)^2 \gg 1$. Gas colliding with the cavity walls, regardless of its history, is diffusely scattered (for rough walls), well mixed, and fully thermalized within a few wall collisions. For the rod at rest in the cavity center then, gas populations infalling from the walls to $S1$ and $S2$ may be taken to be fully thermal and identical in temperature and density. In terms of the velocity distribution functions, this is: $f_I(1,|v|) = f_I(2,|v|)$ and $f_{II}(1,|v|) = f_{II}(2,|v|)$. The velocity distributions for gas infalling from $x = \pm L_c$ are half-Maxwellians, $f_I(j,v)$. When they arrive at $S1$ and $S2$ they are velocity space compressed due to their falls through the gravitational potential, becoming $f_{II}(j,v)$. The rebounding distributions, $f_{III}(j,v)$, are distinct for the two surfaces. After climbing out of the gravitational well, the velocity space expanded distributions $f_{IV}(j,v)$ are rethermalized at the walls. Gravitationally bound gas, $f_V(j,v)$, forms an atmosphere around the rod. The cavity contains blackbody radiation and gas whose mean free path is comparable to or greater than the distance between the rod and the walls. Gas kinetic energy fluxes are much smaller than radiative energy fluxes; in other words, blackbody radiation dominates the system's energy transfers. Small surface temperature variations arising from inelastic collisions are quickly smoothed out by compensating radiative in- or effluxes. This model is valid over a wide range of physically realistic parameters and is well approximated by a planet-sized gravitator in a low density gas housed in blackbody cavity of solar system dimensions. In the following analysis, the rod will be treated one dimensionally; however, it can be shown, in retrospect, that the following results generalize to two and three dimensions.

The net force on the stationary rod can be determined

from conservation of linear momentum, accounting for both incident and reflected particle fluxes. As discussed previously, since $f_I(1,|v|) = f_I(1,|v|)$ and $f_{II}(1,|v|) = f_{II}(2,|v|)$, by symmetry, the net force on the rod (at rest) due to incident gas is zero. However, the net force due to the inelastically reflecting gas need not be zero since $f_{III}(1,|v|) \neq f_{III}(2,|v|)$ and $f_{IV}(1,|v|) \neq f_{IV}(2,|v|)$. Consider the S1 rod end. The incident particle flux density which infalls from the walls at $x = -L_c$ to S1 at $x = -L_g$ is $N_i(1) = \int_0^\infty v f_{II}(1,v)dv$. From conservation of mass, the incident particle flux density is equal to the reflected particle flux density: $N_i(1) = N_f(1) \int_{-\infty}^0 v f_{III}(1,v)\, dv$. The differential momentum flux density for the rebounding gas (taken at $x = -L_g$) is $dF_p(1) = [m_A v]\, dN_f(1) = m_A v^2 f_{III}(1,v)dv$. Only atoms with $v \leq -v_{esc}$ will climb completely out of the gravitational potential well; the remainder will fall back to the rod, form an atmosphere, and eventually evaporate as the $(v \leq - v_{esc})$-tail of $f_V(1,v)$. Accounting for the gravitational back-reaction of the gas on the rod as it climbs out of the gravitational well, the total average steady-state momentum flux density on surface S1 is:

$$F_P(1) \approx M_A \int_{-v_{esc}}^{-\infty} v \sqrt{v^2 - v_{esc}^2}\, [f_{III}(1,v) + f_V(1,v)]\, dv \quad (5)$$

The approximation ($\approx$) is due to the finite cavity size; in the limit of $-L_c \to -\infty$, the expression becomes exact. For S2, $-v_{esc} \to +v_{esc}$ and $-\infty \to +\infty$ in the limits of integration. In the limit of a tenuous atmosphere, the momentum flux density due to the $(|v| \geq |v_{esc}|)$-tail of $f_V(j,v)$ is negligible; in fact, $f_V(j,v)$ is negligible for systems with low gas densities, n_A, and with inelastic response functions, $v_f(j,v_i)$ which do not shift $|v_f|$ significantly below $|v_{esc}|$.

By conservation of linear momentum, the average net momentum flux density (pressure) on the rod as a whole is $\Delta F = F_P(1) - F_P(2)$. If $v_f(1,v_i) \neq v_f(2,v_i)$ in the velocity range of the colliding gas, then except under extremely contrived conditions, one has $\Delta F \neq 0$. In other words, under steady-state thermodynamic conditions, a stationary, gravitating rod with different inelastic responses on its ends can, in principle, experience a non-zero, steady-state force when placed in a suitable gas. If the rod is released, this force can be harnessed to do work at the expense of the heat bath, as discussed previously.

System V. (Solid State) [Author Addition]
The solid state paradox was investigated after the original *JSE* article; details can be found elsewhere.[17] Briefly, it involves the tapping of thermally-generated electrostatic potential energy of p-n junction diodes. It is the first room-temperature, commercially viable second law challenge to appear in the general scientific literature. Laboratory experiments are planned.

Two Broken Symmetries

Each paradox arises due to a synergism between two broken symmetries—one thermodynamic and one geometric. Each is necessary, but alone insufficient. A broken geometric symmetry is constructed into each system. System I possesses almost perfect radial symmetry; this symmetry is broken by the electrical connection from the probe, through the load, to ground. In the case of disconnection, the probe will ran-

domly and radially receive current from the walls through the plasma and radially and randomly return this current to the walls back through the plasma. This is the equilibrium (fully symmetric) case. If the load is connected, however, the probe's return current has an alternate path to ground and the radial symmetry of the current flow is broken. Analogously, in Systems II-IV, the piston's constrained, one-dimensional motion effectively reduces (breaks) the systems' three dimensionality to one.

These broken geometric symmetries are necessary to exploit each system's broken thermodynamic symmetry. The latter may be identified by observing which thermodynamic property, if symmetrized, destroys the paradoxical effect. In System I, the effect is lost if the plasma potential is *symmetrized* to $V_p = 0$. (It is assumed here that for self-emissive probes the floating potential for a probe is equal to the plasma potential.) This can be made zero in several ways including i) ceasing plasma production; ii) achieving the critical plasma density, n_c; or iii) creating a mass-symmetric plasma—a negative ion plasma. More generally, the non-zero V_p can be considered due to either a) the fundamental mass asymmetry between electron and ions; or b) that surfaces preferentially emit electrons or ions depending on values of their surface temperature and work function, and gas ionization potential.

In System II, the paradoxical effect is lost if the work functions of S1 and S2 are equal: $\phi_1 = \phi_2$. Then, the electronic, ionic, and neutral momentum flux densities from all surfaces are identical, rendering zero the pressure differential between piston faces. In general, the symmetry condition, $\phi_1 = \phi_2$, is difficult to achieve unless S1 and S2 are the same material—a trivial case.

In System III, the effect is lost if the desorption rate ratios for S1 and S2 are equal: $\alpha_1 = \alpha_2$. As seen from Equation 3, this requires either fine tuning in values of surface density, partition functions, and desorption energies, or that S1 and S2 be identical substances. As with ϕ in System II, the symmetry condition, $\alpha_1 = \alpha_2$, is difficult to achieve unless S1 and S2 are identical. In System IV, the effect is lost if $v_f(1,v_i) = v_f(2,v_i)$. This is most easily accomplished by symmetrizing the rod's composition.

Each broken thermodynamic symmetry (in V_p, ϕ, α, or $v_f(j,v_i)$) occurs naturally under either equilibrium or non-equilibrium conditions and allows momentum flux asymmetries to arise. Via the broken geometry symmetry, the broken thermodynamic symmetry is exploited to do work. Both broken symmetries appear to be necessary since the thermodynamic quantities V_p, ϕ, α, $v_f(j,v_i)$ are spatially homogeneous (independent of spatial variables); therefore, by themselves they are insufficient to direct momentum fluxes to do work. This requires the broken spatial (geometric) symmetry; in System I it is accomplished by an electrical conductor and in Systems II-IV by a piston. From these four examples,[1] a conjecture is induced: Given a spatially homogeneous thermodynamic property that causes an asymmetric momentum flux (under equilibrium or non-equilibrium conditions), a second broken geometric symmetry is necessary and, if suitably arranged, can be sufficient to do work solely at the expense of a heat bath in violation of the second law.

(It is noted that System IV was not used in inducing this conjecture; on the contrary, it was deduced from it. As a personal challenge, the author cooked it up to test whether the conjecture induced from Systems I-III could hold for gravi-

159

tational systems as it seemed to apply for electromagnetic ones. Apparently, it can.)

[Author Note: Unknown to the author at the time of the original *JSE* article, several years prior (1992) a proof was given by Zhang and Zhang,[47] demonstrating formally much of the essence of this last section. Specifically, they showed that what they called "spontaneous momentum flows" (SMF) are necessary and sufficient conditions for second law violations. These paradoxes can be shown to satisfy the Zhang criteria for second law violation. In the last year it has also been realized that the entire class of paradoxes share the common feature of macroscopic potential gradients (*e.g.*, chemical, electrostatic, gravitational) by which particle fluxes are used to extract energy ultimately from a heat bath (Ref. 13, p. 195).]

Acknowledgements

This work was supported by a 1995 University of San Diego (USD) Faculty Research grant, and a 1996-97 USD University Professorship and a 1997 NASA-ASEE Faculty Fellowship. The author thanks Drs. William F. Sheehan and Jack Opdycke for illuminating discussions.

References

1. Sheehan, D.P. 1998. "Four Paradoxes Involving the Second Law of Thermodynamics," *J. Sci. Explor.*, 12, 2, 303-314.
2. Sheehan, D.P. 1995. "A Paradox Involving the Second Law of Thermodynamics," *Phys. Plasmas*, 2, 1893.
3. Sheehan, D.P. and Means, J.D. 1998. "Minimum Requirement for Second Law Violation: A Paradox Revisited," *Phys. Plasmas*, 5, 2469-2471.
4. Capek, V. and Sheehan, D.P. 2002. "Quantum Mechanical Model of a Plasma System: A Challenge to the Second Law of Thermodynamics," *Physica A*, 304, 461.
5. Sheehan, D.P. 1996. "Another Paradox Involving the Second Law of Thermodynamics," *Phys. Plasmas*, 3, 104.
6. Sheehan, D.P. 1998. "Dynamically-maintained, Steady-state Pressure Gradients," *Phys. Rev. E*, 57, 6660-6666.
7. Sheehan, D.P. 2000. Reply to "Comment on 'Dynamically-maintained Steady-state Pressure Gradients,'" *Phys. Rev. E*, 61, 4662.
8. Sheehan, D.P. 2001. "The Second Law and Chemically-induced, Steady-state Pressure Gradients: Controversy, Corroboration, and Caveats," *Phys. Lett. A*, 280, 185.
9. Sheehan, D.P., Glick, J., and Means, J.D. 2000. "Steady-state Work by an Asymmetrically Inelastic Gravitator in a Gas: A Second Law Paradox," *Found. Phys.*, 30, 1227.
10. Sheehan, D.P. and Glick, J. 2000. "Gravitationally-induced, Dynamically-maintained, Steady-state Pressure Gradients," *Phys. Script.*, 61, 635.
11. Sheehan, D.P., Glick, J., Duncan, T., Langton, J.A., Gagliardi, M.J., and Tobe, R. 2002. "Phase Space Portraits of an Unresolved Maxwell Demon," *Found. Phys.*, 32, 441.
12. Sheehan, D.P., Glick, J., Duncan, T., Langton, J.A., Gagliardi, M.J., and Tobe, R. 2002. "Phase Space Analysis of a Gravitationally-induced, Steady-state Nonequilibrium," *Phys. Scripta*, 65, 430.
13. Sheehan, D.P., ed. 2002. *Proceedings of First International Conference on Quantum Limits to the Second Law*, San Diego, CA, July, AIP, Vol. 643 (American Institute of Physics,

Melville, NY).
14. Capek, V.C. and Sheehan, D.P. 2004. *Challenges to the Second Law of Thermodynamics*, Frontiers of Physics Series, Kluwer Academic, Amsterdam.
15. Leff, H.S. and Rex, A.F., eds. 1990. *Maxwell's Demon: Entropy, Information, Computing*, Princeton Series in Physics, Princeton.
16. Leff, H.S. and Rex, A.F., eds. 2003. *Maxwell's Demon 2: Entropy, Classical and Quantum Information*, Computing, Institute of Physics, Bristol.
17. Sheehan, D.P., Putnam, A., and Wright, J.H. 2002. "A Solid-state Maxwell Demon," *Found Phys.*, 32, 1557.
18. Allahverdyan, A.E. and Nieuwenhuizen, Th.M. 2000. "Optimizing the Classical Heat Engine," *Phys. Rev. Lett.*, 85, 232-235.
19. Allahverdyan, A.E. and Nieuwenhuizen, Th.M. 2000. "Extraction of Work from a Single Thermal Bath in Quantum Regime," *Phys. Rev. Lett.*, 85, 1799-1802.
20. Allahverdyan, A.E. and Nieuwenhuizen, Th.M. 2000. "Steady Adiabatic State: Its Thermodynamics, Entropy Production, Energy Dissipation, and Violation of Onsager Relations," *Phys. Rev. E*, 62, 845-850.
21. Capek V. and Frege, O. 2000. "Dynamical Trapping of Particles as a Challenge to Statistical Thermodynamics," *Czech. J. Phys.*, 50, 405-423.
22. Capek, V. and Mancal, T. 1999. "Isothermal Maxwell Demon as a Molecular Rectifier," *Europhys. Lett.*, 48, 4, 365-371.
23. Capek, V. and Tributsch, H. 1999. "Particle (electron, proton) Transfer and Self-organization in Active Thermodynamic Reservoirs," *J. Phys. Chem. B*, 103, 3711-3719.
24. Capek, V. 1998. "Isothermal Maxwell Demon as a Quantum 'Sewing Machine,'" *Phys. Rev. E*, 57, 3846-3852.
25. Capek, V. and Bok, J. 1998. "Isothermal Maxwell Demon: Numerical Results in a Simplified Model," *J. Phys. A: Math Gen.*, 31, 8745-8756.
26. Capek, V. 1997. "Isothermal Maxwell Demon," *Czech. J. Phys.*, 47, 845-849.
27. Capek, V. 1998. "Isothermal Maxwell Demon: Swing (fish-trap) Model of Particle Pumping," *Czech. J. Phys.*, 48, 879-901.
28. Capek, V. 1997. "Isothermal Maxwell Demon and Active Binding of Pairs of Particles," *J. Phys. A: Math. Gen.*, 30, 5245-5258.
29. Capek, V. and Bok, J. 1999. "A Thought Construction of Working Perpetuum Mobile of the Second Kind," *Czech. J. Phys.*, 49, 1645-1652.
30. Capek, V. 2001. "Twilight of a Dogma of Statistical Thermodynamics," *Mol. Cryst. Liq. Cryst.*, 355, 13-23.
31. Gordon, L.G.M. 1983. "Maxwell's Demon and Detailed Balancing," *Found. Phys.*, 13, 989-997.
32. Gordon, L.G.M. 1981. "Brownian Movement and Microscopic Irreversibility," *Found. Phys.*, 11, 103-113.
33. Liboff, R.L. 1997. "Maxwell's Demon and the Second Law of Thermodynamics," *Found. Phys. Lett.*, 10, 89-92.
34. Nikulov, A.V. 2001. "Quantum Force in a Superconductor," *Phys. Rev. B*, 64.
35. Nikulov, A.V. and Zhilyaev, I.N. 1998. "The Little-Parks Effect in an Inhomogeneous Superconducting Ring," *J. Low Temp. Phys.*, 112, 227-235.
36. Motley, R. 1975. *Q-Machines*, Academic Press, New York, pp. 11-14.
37. Rynn, N. and D'Angelo, N. 1960. *Rev. Sci. Instrum.*, 31, 1326.

Appendix H

Is Loschmidt's Greatest Discovery Still Waiting for its Discovery?

By Andreas Trupp (Reprinted by permission of the author)

Abstract

In 1868 J.C. Maxwell proved that a perpetual motion machine of the second kind would become possible, if the equilibrium temperature in a vertical column of gas subject to gravity were a function of height. However, Maxwell had claimed that the temperature had to be the same at all points of the column. So did Boltzmann. Their opponent was Loschmidt, who died more than 100 years ago in 1895. He claimed that the equilibrium temperature declined with height, and that a perpetual motion machine of the second kind operating by means of such column was compatible with the second law of thermodynamics. Thus he was convinced he had detected a never ending source of usable energy for mankind. In this article, new proof is given for the hypothesis that the equilibrium temperature is indeed a function of height.

Moreover, the consequences (of the feasibility of a Perpetual Motion Machine of the Second Kind) for the nature of time are discussed: The fact that water in a cup, when stirred, will climb the walls of the cup as a result of the centrifugal "force" at work, is -- as General Relativity tells us -- a result of the special way the distant stellar masses of the universe are distributed. Likewise, the fact that cold water, when added to a cup containing hot coffee, will mix with the coffee to form a liquid of uniform temperature, has to be a result of the special state of matter in the universe billions of years ago, since it could no longer be regarded as a consequence of the Second Law. With the Second Law dismissed as false, only the initial state of things, at a given point in time, can explain why an increase or a decrease in order can be observed. The often discussed paradox, that is the question why temperature differences within an ideal gas will always vanish though all motions of the molecules are reversible (so that increases in temperature differences should be as frequent as reductions of these differences) is thereby resolved: The initial state of the gas to start with (which itself is causally dependent on prior states of things) isn't of the right kind for generating temperature differences. This recognition gives rise to revisit Boltzmann's famous dispute with Zermelo. In a universe endless both in time and in space, he argues, there must exist "islands" in which, by random processes, matter is organized, whereas the universe is barren and at uniform temperature elsewhere. Living beings (including intelligent machines) on such an island will *define* the arrow of time by saying that the future is the *less* organized state, while the past is the *more* organized state of their island. As long as there are no "islands within the islands," that is to say: as long as there aren't too many perpetual motion machines of the second kind at work, this definition is quite unambiguous. The universe is hence in possession of different states, but is lacking of an intrinsic ordering of these states by the category earlier / later. Instead, such an ordering is extrinsic. It seems that Boltzmann's view of the arrow of time is quite correct despite the fact that cosmology, by assuming the Big-Bang at the "origin" of the universe, is dismissing the assumption of a universe endless both in space and time.

Introduction

When taking a look at any textbook on general physics, one finds the second law of thermodynamics formulated in two equivalent ways: "The total entropy of an isolated system can never decrease", and "A perpetual motion machine of the second kind is impossible." Today, doubting the impossibility of such a machine is just as inconceivable as is the assertion that a perpetual motion machine of the FIRST kind might exist, which creates energy from nothing. A closer investigation, however, reveals that in the second half of the nineteenth century, a vivid debate was held among most reputable scientists on the possibility of a perpetual motion machine of the second kind. One of the most prominent propagators in favor of such a possibility was Josef Loschmidt, a name today well known even to school-kids through the celebrated Loschmidt's number. A perpetual motion machine of the second kind would be capable of permanently creating, in a cycle, mechanical energy from just one single reservoir of heat. Thus it would become possible, for instance, to convert the (dissipated) energy of heat contained in the air into mechanical energy without requiring a second, colder reservoir for the absorbtion of the refuse heat. Therefore mankind would have available a source of energy that practically cannot be exhausted.

Loschmidt and Maxwell asserted that, if there were a stratification of temperature in a column of gas subject to gravity, the construction of a perpetual motion machine of the second kind would be possible. Up to the present day, no one has ever challenged that assertion. Maxwell believed that the temperature of the gas subject to gravity could not be stratified, but had to be the same at all points. He did not provide a special proof; rather, he intuitively extended his formula of velocities of molecules (which had been derived without regarding gravity) to a gas subject to gravity. Boltzmann sided with Maxwell; in contrast to Maxwell, he attempted to prove that the homogeneous temperature of a gas subject to gravity was ensured by the kinetic theory of gases. Loschmidt, however, was convinced that a perpetual motion machine of the second kind was compatible with the second law of thermodynamics. In that point, he disagreed with Clausius, Thomson, Boltzmann, and Maxwell. In particular, he believed that a perpetual motion machine of the second kind could be operated by means of a vertical column of gas, the temperature of which he claimed to be stratified.

In the 20th century, Loschmidt's "revolutionary" assertion has hardly been paid any attention. After all, it was mentioned by Stephen G. Brush in his 1978 book: "The Temperature of History." However, Brush does not give more than a clue when telling his readers that the dispute over the stratification of temperature between Boltzmann and Loschmidt provided a contribution to the debate on the second law of thermodynamics. No further details are offered.

In recent times (1995), it was Claude Garrod [1] who tried to give a new proof of the uniformity of temperature. His arguments will be scrutinized further below.

On the history of the second law of thermodynamics

Though the expression "perpetual motion machine of the second kind" was only introduced by Ostwald towards the end of the 19th century, the impossibility of such machine had been postulated as an axiom by Clausius already in 1849 and by Thomson in 1850. Both Clausius and Thomson are considered the discoverers of the second law of thermodynamics. [1a]

One may wonder how Clausius and Thomson could obtain their firm belief in the truth of their axiom. The fact alone that such a machine had not been invented until those days is not capable of explaining this conviction. In addition, one has to take into account that already in the 18th century the opinion of the impossibility of a perpetual motion machine prevailed, long before the theorem of the conservation of energy or the distinction between the first and the second law of thermodynamics were advanced. The idea of a system of movable parts that, having come to rest once, would still be able to get into motion on its own, was simply inconceivable [2]. Of course, the 18th century scientists were not yet familiar with the kinetic theory of heat and did not realize that apparent rest turns into motion in the microscopic perspective. Such knowledge would have impeded the formation of Clausius' and Thomson's axiom.

Picking up reflections previously published by Carnot, Thomson declared the impossibility of a perpetual motion machine the foundation of his further investigations in the field of thermodynamics: *"It is impossible, by means of inanimate material agency, to derive mechanical effect from any portion of matter by cooling it below the temperature of the coldest surrounding objects."* [3] From the impossibility of such a perpetual motion machine one can easily infer the second law in a very general form, stating that a system will not depart from an attained state of equilibrium without interference from outside. Loschmidt accepted the second law in such a general form only (more precisely: he believed that the second law could be derived from the mechanical principle of least action). However, as emphasized by Loschmidt several times, it is impossible to invert the order of inference, i.e. it is not permitted to infer the impossibility of a perpetual motion machine of the second kind from the second law in its very general form:

"From these reflections one can draw the conclusion that the second law of thermodynamics can be inferred from the axiom of Clausius 'It is impossible to transfer heat from a colder to a warmer body without compensation', or from the equivalent one of W. Thomson 'It is impossible, by means of inanimate material agency, to derive mechanical effect from any portion of matter by cooling it below the temperature of the coldest of the surrounding objects', that the inversion of that inference, however, is not permissible, because the content of the second law is more general than that of those axioms." [4]

Maxwell's (hypothetical) perpetual motion machine of the second kind

In the 1860's and 70's scientists dealt with the question of whether a gas, which is subject to gravity in an insulated column, had the same temperature at all points, or whether its temperature was a function of height.

Maxwell was convinced that the temperature of a gas subject to gravity had to be uniform at all heights. Moreover, like Thomson and

Clausius, he regarded a perpetual motion machine of the second kind to be impossible. However, he was very sure that if there were a gradation of temperature and if that gradation were different for different substances, a perpetual motion machine would be possible:

"In fact, if the temperature of any substance, when in thermic equilibrium, is a function of the height, that of any other substance must be the same function of the height. For if not, let equal columns of the two substances be enclosed in cylinders impermeable to heat, and put in thermal communication at the bottom. If, when in thermal equilibrium, the tops of the two columns are at different temperatures, an engine might be worked by taking heat from the hotter and giving it up to the cooler, and the refuse heat would circulate round the system till it was all converted into mechanical energy, which is a contradiction to the second law of thermodynamics. The result as now given is, that temperature in gases, when in thermal equilibrium, is independent of height, and it follows from what has been said that temperature is independent of height in all other substances." [5]

Modifying the device introduced by Maxwell, we put up with one column (filled with gas) only, which is thermally isolated from its surroundings (see fig. 1). A vertical pipe is arranged in a way that it penetrates the entire column from its bottom to its top. The material of the pipe consists of small sections permeable to heat; each of such sections is followed by a section that is not permeable to heat. So the number of permeable sections equals the number of impermeable ones. In the interior of that vertical pipe, a piece of metal (of cylinder shape) suspended by a rope is allowed to move up and down. At the top of the column, that metal piece has the temperature of the upper part of the column. Initially (before the device begins its operation), that temperature is also the temperature of the ambient. But when being slowly lowered to the floor, the metal piece adopts the temperature of the gas surrounding it; it gathers tiny bits of heat at every height by getting into thermal contact with the permeable sections. Thus it becomes hotter (without disturbing the phenomenon of a temperature gradation of the gas as such).

Finally it has reached the bottom (and the maximum temperature). Now it is towed back to the top in such a fast way that the temperature of the piece of metal has no chance of declining. In order to move the piece of metal up and down, no net work has to be spent: the expenditure of work when making it move upwards it completely compensated by the gain in work when it is lowered to the bottom. The hot metal piece, however, can make water boil, and by the steam thus created a steam engine can be run. The refuse heat of that engine flows into the upper part of the column (not into the ambient).

One might be tempted to assume that the process of creating work has to come to a standstill as soon as the temperature of the column has reached uniformity (due to the extraction of heat at its bottom and the adding of refuse heat at its top). However, the starting point of our (and Maxwell's) reflections was the hypothetical assumption (which Maxwell did not believe to be true in reality) that a uniform temperature of a column of gas subject to gravity is NOT A STATE OF EQUILIBRIUM. Hence we have to conclude that the gas, left to itself during that break in work creation, will resume its state of temperature gradation. Then, the whole process can start again. The internal energy of the gas as a whole will thus be diminished and turned into mechanical work without a second heat reservoir.

Strictly speaking, Maxwell's original device only demonstrates that DIFFERING temperature gradients of two substances enable the construction of a perpetual motion machine of the second kind, whereas the modification gives proof of the possibility of such machine already in the event of a temperature gradation existent in one single substance, that is in case of a temperature gradation as such.

A proof of the stratification of temperature in gases subject to gravity (confined to ideal gases)

I. Imagine an ideal gas in a box. Let us assume those molecules were capable of moving in one direction only, that is in the vertical direction. In order to assure this, they find themselves in vertical columns. The diameter of each of the many vertical columns is just that of the spherical molecule itself. In addition to their vertical component of velocity **v,** the molecules can store some energy by rotating around an axis. They thus have two degrees of freedom. Everytime two molecules collide with each other within a vertical column, it shall be a matter of chance how the rotational motion of the two collision partners will be affected thereby. For instance, given that the vertical components of motion of the two molecules are equal in amount, but of opposite directions, both molecules will bring their vertical velocity down to zero, having their previous energy of vertical motion transferred to their energy of rotational motion. Both the translational momentum and the angular (rotational) momentum of the ensemble would be conserved in this process: If, for instance, the translational momentum of the first molecule was $+mv\_1$, while the momentum of the second molecule was $-mv\_1$, the total momentum of the ensemble is zero. This momentum is conserved even after the vertical velocities of both molecules have reduced to zero. Furthermore, let us assume that the *angular* momentum of the first molecule was **+I omega_1**, while that of the second molecule was **-I omega_1.** Then the total angular momentum was zero. If, after the collision, both molecules have increased their angular momentum by an equal amount (so that the total loss of the energy of *translational* motion is fully compensated by the increase in kinetic energy of *rotational* motion), the total angular momentum of the two molecules is still zero.

Moreover, since the molecules are capable of central collisions only, we could as well say that the collisions as such do not affect their translational motions at all, if only we decide to switch the identities of the collision partners. In this picture, they penetrate each other as if they were not of tangible matter. We may thus ascribe green T-shirts to all molecules rising upward, and red T-shirts to all molecules moving downward, with the T-shirt being passed to the collision partner like a baton in a relais-race.

Hence, at each encounter with another molecule, the total energy of a *single* molecule may change, and so may its vertical velocity, and this change in vertical velocity would be brought about at the expense of its own rotational energy and/or that of its collision partner. Now, everytime the rising molecule is getting a push upward at the expense of the rotational energy, it will, thereafter, lose this extra push (or of a fraction thereof) by the effect of gravity on ist further way upward. Then, however, it is obvious that the gain in height was achieved not solely at the expense of the translational, but also of the rotational energy.

Is this expense of rotational energy (that took place to the benefit of the *potential* energy) being „refunded" by the potential energy, if, thereafter, the rising molecule *reduces* its vertical velocity as a result of *another* collision? The answer is in the negative: This gain in rotational energy is not at the expense of the *potential* energy, but of the energy of translational, vertical motion. The „refund" by the recipient, that is by the potential energy, is not granted until the molecule starts its way downward (with a red T-shirt on); a previous gain in vertical velocity brought about by the action of gravity on the molecule's downward journey might then be taken away from the molecule as a result of a collision, and this loss would then be to the benefit of the rotational energy.

That is to say: The gain in potential energy of the molecules wearing green T-shirts is at the expense of the kinetic energy stored in *both* of the two degrees of freedom.

II. Let us now switch to an ideal gas in which the molecules are capable of motions in three dimensions, with their rotational energy being vanshingly small. During each temporal unit, molecules of different total energies v_0 (ranging from zero to infinity) are rising from the bottom of the box (the index 0 stands for the bottom of the box). Velocity, in this context, means initial total velocity comprising both vertical and lateral components. Each initial velocity (that a molecule might be in possession of) is assigned to a certain velocity class v_k (with **k** ranging from 1 to infinity) that comprises velocities between v_k and $v_k + dv_0$. Let us imagine that each molecule is displaying, on its T-shirt, its initial class of velocity to which it belongs.

In most cases in which a collision takes place, the total energy of the molecule is changed (while the total energy of the ensemble of the two molecules is conserved). As a consequence of such a sudden change in energy, the identity of the molecule shall be switched with that of another molecule, which, at the same height above the bottom, is in possession of a total energy equal to the total energy that our molecule was in possession of just prior to the collision. At each height, a whole class of molecules (being in possession of the right amount of total energy) will be available to choose from. Their vertical upward velocities range from zero to the maximum velocity at which all the energy is stored in that component of their total energy. Likewise, their lateral energies range from zero to the maximum energy (which is the total energy of the class). Moreover, when a rising molecule is undergoing a collision that reverses its path in a way that it would start a downward motion, it switches identities with its collision partner (as was the case in the preceding model).

Let us again assume that all molecules bound for the ceiling wear green T-shirts, whereas those bound for the bottom are wearing red T-shirts. Thus a molecule with a green T-shirt is able to reach the height **h**, if its initial total energy (comprising three degrees of freedom) was $(2gh)^{1/2}$, with **g** denoting the gravitational acceleration it is subject to. Thus the gain in height of a molecule is at the expense of all the components of its energy, as was the case in the preceding model. The contribution of the lateral components (of the kinetic energy) to the gain in potential energy of a molecule, brought about in a collision, is not refunded by the recipient, that is the potential energy of the molecule, until it changes the color of its T-shirt to red (indicating its downward motion).

If the initial total energy of a molecule was not high enough to make it to the ceiling, the color of the T-shirt is changed at its climax. In a state of equilibrium, the number green T-shirts equals the number of red T-shirts.

III. We shall now be interested in getting to know the average square of the molecules' vertical velocity (that have risen from the bottom) once they have reached the height **h**. When penetrating a thought horizontal surface at that height (or when being reflected there), the average square of their velocity is determined in the following way: At height **h**, the vertical velocities $v_h=(v_0^2-2gh)^{1/2}$ of the molecules passing through the thought surface range from zero to infinity. These velocities v_h are being distributed among a large number of classes. For each velocity class, the number of molecules that belong to that class is being determined. For this purpose, all the molecules penetrating the thought surface at height **h** during a temporal interval of unit length are subject to scrutiny. The numbers of molecules belonging to each velocity class are then multiplied by the respective square of their velocity. The so formed products are added up. The total sum is divided by the total number of molecules that penetrated the thought surface during the temporal interval of unit length.

In mathematical terms, we get:

(1)

$$T_h = K \, \overline{v_h^2} = K \, \frac{\int_{\sqrt{2gh}}^{\infty} (v_0^2 - 2gh) \, N_0 \, f(v_0) \, dv_0}{\int_{\sqrt{2gh}}^{\infty} N_0 \, f(v_0) \, dv_0}$$

T_h denotes the temperature of the gas at height **h**, which presents itself as the product of the mean square of the velocity at that height (v_h) and a constant **K**. The second constant N_0 denotes the total number of molecules rising from the ground during a temporal interval of unit length. The function **f(v_0)** is designed in a way that — due to another constant **C** which is part of that function — an integration from zero to infinity yields unity (dimensionless). Moreover, it tells us about the distribution of the absolute velocities of those molecules which rise from the ground during each second. For a given velocity class that comprises a certain number of molecules, it is clear that the arrival rate (molecules per second) at the ceiling (at height **h**) is identical to the departure rate from the ground, even if the time needed for the ascent varies from molecule to molecule. This is a necessary consequence of the state of equilibrium, which was presupposed to exist.

That function f(v0) is identical with Maxwell's distribution of total velocities. Since the ordinary Maxellian formula is about the distribution of velocities within a unit volume, and not (as we would like to have it) about the distribution of velocities at which the molecules rise from the ground, the Maxwellian formula we use is derived from the ordinary Maxwellian distribution by adding another v0 as a factor (and by adjusting the norming constant): The faster the molecules within a layer just above the ground, the more often they will hit the ground, where -in our model-they will face an immediate reflection. Hence (1) turns into (2)

$$T_h = K \overline{v_h^2} = K \frac{\int_{\sqrt{2gh}}^{\infty} (v_0^2 - 2gh) \, N_0 \, C \, v_0^3 \, e^{-av_0^2} \, dv_0}{\int_{\sqrt{2gh}}^{\infty} N_0 \, C \, v_0^3 \, e^{-av_0^2} \, dv_0}$$

The constant **a** equals $(2RT_0)^{-1}$, with T_0 denoting the temperature of the gas at the bottom, with **R** denoting the gas constant. The term **g** denotes the gravitational acceleration. Both integrals can be developed by a succession of partial integrations, according to the well known rule: (3)

$$\int u(v_0) \, w'(v_0) \, dv_0 = [uw] - \int wu' \, dv_0$$

The function **w'** can, in each partial integration, be expressed as:

$$v_0 \, e^{-av_0^2}$$

For **w**, we get:

$$-\frac{1}{2a} \, e^{-av_0^2}$$

As a result, we find: (4)

$$T_h = K \overline{v_h^2} = K \frac{\dfrac{gh}{a} + \dfrac{1}{a^2}}{gh + \dfrac{1}{2a}}$$

In order to determine the constant **K**, the term **gh** is set to zero. Then Equation (4) yields: $T_h = 2K/a = 4KRT_0$. Because of $T_h = T_0$, we get $K = (4R)^{-1}$. The fact that the average squared velocity is thus **4RT** -rather than **3RT** as given in textbooks for the molecules scattered in space- is accounted for by realizing that the average kinetic energy of the molecules in a unit volume of space cannot be identical with the average kinetic energy at which these molecules penetrate a thought surface, since -at a given distribution of velocities within a unit volume of space- the molecules with higher speeds manage to reach that surface more often than those molecules with lower speeds.

Hence (4) turns into

(5)

$$T_h = \frac{ghT_0 + 2RT_0^2}{2gh + 2RT_0} = \frac{T_0 + \dfrac{2RT_0^2}{gh}}{2 + \dfrac{2RT_0}{gh}}$$

For large values of **h**, T_h approaches **1/2 T_0** . There is a decline in temperature with height, but this decline is not linear. For a temperature of 300 ° Kelvin, and a difference in height of 100 meters, the decline in temperature for air is about 1.4 degrees. As this gradient is overadiabatic (the adiabatic gradient being 1 deg per 100m), it will not be observed in a real gas, since convection will prevent its full materialization. Instead, the observed gradient will be almost identical with the adiabatic one (disregarding heat radiation).

III. If, different from our previous assumption, the gain in height of the molecules were at the expense of the vertical component of velocity only, f(v0) would be identical with the Maxwellian distribution of velocities in the one-dimensional gas (in which v0 denotes the vertical component of velocity, not the total velocity). The temperature gradient would be quite different then. We would get:

(6)

$$T_h = K\,\overline{v_h^2} = K\,\frac{\displaystyle\int_{\sqrt{2gh}}^{\infty} (v_0^2 - 2gh)\, N_0\, C\, v_0\, e^{-a v_0^2}\, dv_0}{\displaystyle\int_{\sqrt{2gh}}^{\infty} (v_0^2 - 2gh)\, N_0\, C\, e^{-a v_0^2}\, dv_0} = K\, 2RT_0$$

The temperature would hence be independent of height, or : $T_0 = T_h$. This explains why, in recent times, Walton [6] and also Garrod [1] "proved" the uniformity of temperature in a vertical column of gas subject to gravity, when, in their models, they assumed that no mutual collisions of the molecules would take place, so that the gain in height would have to be at the expense of the *vertical* component of velocity only (the two lateral components of motion are not affected by the rise of the molecules). This would undoubtedly be the case if the walls of our container were not rugged, but smooth (see also *F.L. Roman, J.A. White, S.Velasco,* Microcanonical single-particle distributions for an ideal gas in a gravitational field, Eur. J. Phys., vol. 16 -1995-, pp. 83-90, additional remarks in Eur.J. Phys., vol. 17 -1996-, pp. 43-44; *Charles A. Coombes, Hans Laue*, A paradox concerning the temperature distribution of a gas in a gravitational field, Am. J. Phys., vol. 53 -1985-, pp. 272-273).

So, which of the two distributions matches with reality? Imagine the molecules were extensionless (zero in diameter), but would nevertheless be capable of undergoing mutual collisions. Then all collisions would be *central* collisions. However, in a central collision of two molecules of the same mass, the two velocities (taken as vectors) are just mutually exchanged: If we chose a reference system in which one of the two molecules is at rest, the second molecule would just push the first molecule away, and would come to rest at exactly the same spot that, so far, has been occupied by the first molecule. As a consequence, we could decide to switch the identities of the two molecules, and could say that they penetrated each other as if they were not of tangible matter. Such molecules are unable to increase their vertical component of velocity (or any of the other two components of velocity) by means of collisions. Thus, when their vertical component of velocity is exhausted, they have to give up their upward journey. This would be their fate despite the fact that most of them still have lateral components of velocity greater than zero, which, however, they could not convert into a vertical component.

However, a molecule *would* clearly be able to increase its vertical component of velocity by collisions if (and only if) a collision is *not* central. Imagine that the mass of the molecule were concentrated in the center of a ping-pong ball (the mass of which shall be close to zero). Let us again choose the reference system in a way that this molecule is at rest. A second molecule could push away our molecule in a non-central collision, but would be unable to take over the role the first molecule has played so far. That is to say: Even with a switch of identities, a change in the vertical component of velocity (as a result of the collision) cannot be excluded (though the sum of the two momenta is conserved). In case the ping-pong ball had a definite mass, it could even store kinetic energy by rotating, and could transform this rotational kinetic energy into translational kinetic energy in another collision.

Moreover, the Maxwell-Boltzmann distribution of velocities (which was applied in the foregoing) is based on the assumption that, for a molecule facing its next collision, it is a pure matter of chance at which velocity (more precisely: in which unit cell of velocity space) it will find itself thereafter. Only one single restriction is being imposed: The total energy of all molecules has to stay invariant. In the words of J.C. Maxwell (Theory of Heat): "*The first thing we must notice about this moving system is that even if all the molecules have the same velocity originally, their encounters will produce an inequality of velocity, and that this distribution of velocity will go on continually. Every molecule will then change both its direction and its velocity at every encounter; and, as we are not supposed to keep a record of the exact particulars of every encounter, these changes of motion must appear to us very irregular if we follow the course of a single molecule. If, however, we adopt a statistical view of the system, and distribute the molecules into groups, according to the velocity with which at a given instant they happen to be moving, we shall observe a regularity of a new kind in the proportions of the whole number of molecules which fall into each of these groups.*" It is quite obvious that Maxwell is assuming the molecules **not** to be confined to *central* collisions only, otherwise it would not be true to say that, in case the molecules have the same velocity originally, their encounters will produce an inequality of velocity. For, if central collisions were the only collisions to occur, we could adopt the view according to which the molecules penetrate each other as if they were not of tangible matter, thereby making it evident that the molecules do *not* have the means for changing their initial "distribution" of velocities (in which all molecules have the same absolute velocity).

Since the Boltzmann-Maxwell distribution law is a good match with reality, all components of a molecule's energy seem to be capable of undergoing changes in a collision, quite different from the competing assumption saying that only *central* collisions are allowed to occur.

A refutation of Boltzmann's "proof" of a uniform temperature of gases subject to gravity

The controversy on the distribution of temperature in such gas was held most vividly by Loschmidt and Boltzmann. Loschmidt postulated the decline of temperature with height (both for solid substances and gases) and therefore the possibility of a perpetual motion machine of the second kind: "*The result achieved in the foregoing, according to which local differences in temperature can occur and persist in a system of aligned atoms exposed to gravity, is in direct contradiction to a basic theorem of the theory of heat, which claims that no work can be obtained from the active force of a system of bodies being in the state of thermal equilibrium."* [7] As opposed to this, Boltzmann asserted that the gas had a homogeneous temperature. It is worth mentioning that Boltzmann refrained from attacking Loschmidt with the negative argument that the gradation of temperature entailed the possibility of a perpetual motion machine of the second kind.

Boltzmann explicated his view by combining his distribution of velocities (which had been set up for an isothermal gas beyond the reach of gravity) and the formula of the barometric gradient.[8] Thus he obtains the following distribution of velocities:

(15)

$$dN = C \, e^{[-h(u^2+v^2+w^2)-kz]} \, du \, dv \, dw \, dz$$

The two horizontal components of velocity are represented by **u** and **v, w** stands for the vertical component of velocity. The variable **z** measures the height of a layer on a vertical axis of coordinates; **C, h** and **k** are defined as (unknown) constants. The expression **dN** denotes that portion of molecules (their total number being **N**) the velocities of which range from **u** to **u + du,** from **v** to **v + dv,** from **w** to **w + dw,** and the **z** - coordinates of which range from **z** to **z + dz.** Boltzmann conceives a gas in which the distribution mentioned above is established at time zero. He asks whether or not the distribution is about to change in the course of time as a result of gravity; in other words, whether or not equation (15) is still capable of describing the state of the gas after some time has passed. In case the answer is in the positive, he regards his thesis (of a homogeneous temperature of the gas) as being approved. For the sake of simplicity, he assumes (for a while) that no collisions of molecules take place. The only collisions they undergo are the reflections by the vertical walls of the container. Since these reflections do not have any effect on the horizontal components of their velocities (in absolute numbers), it is clear that the number of those molecules the components of which ranged from **u** to **u + du** and from **v** to **v + dv** at time zero is not affected by gravity and hence remains constant over time.

With regard to that number, he asks whether the fraction of molecules the vertical velocity **w** of which ranged from **w** to **w + dw** , and the **z**-coordinates of which ranged from **z** to **z + dz** at time zero, remain in **dw**- and **dz**-intervals of the same widths even after the molecules have been subject to gravitation. As all the molecules, in a given time, increase (or decrease) their vertical velocities by the same amount as a result of gravity, the width of the w-interval, that is **dw,** does indeed remain unchanged.

Boltzmann's computations, however, lead to the outcome that the width of the z-interval, i.e. **dz**, does not stay constant over time. This is quite evident: the molecules have different velocities ranging within the interval **dw** at time zero, hence their mutual distance is changing in the course of time. The widening of the **dz** – interval does not vanish when it comes to differential values of **t**. Let us see why: The molecule with the smallest initial value of **z** and the smallest initial value of **w** will, after time interval **dt** has elapsed, find itself at

(16)

$$z + (w + 1/2 \, g \, dt)dt$$

The molecule with the greatest initial value of z (which is z + dz) and the greatest initial value of w (which is w + dw) will, after time interval t has elapsed, find itself at

(17)

$$z + dz\_0 + (w + dw + 1/2 \, g \, dt)dt$$

Subtracting the first expression from the second one, we arrive at

(18)

$$dz\_0 + dw \, dt$$

for the increased width of the z-interval. The initial **z** -interval would remain unchanged if not only **dt,** but also **dw** approached zero. The error would then be of second order only, and even if those errors were summed up during an integration, they would still vanish as being equivalent   to

(19)

$$dwdt + dwdt + \ldots = dwdt \, \frac{\Delta t}{dt} = dw \, \Delta t$$

and hence to zero. The interval **dw**, however, cannot shrink to zero for statistical reasons: The probability that a single, depicted molecule lies outside an interval **dw** amounts to:

(20)

$$P = \frac{\dfrac{\Delta w}{dw} - 1}{\dfrac{\Delta w}{dw}} = 1 - \frac{dw}{\Delta w}$$

The probability of ALL molecules lying outside **dw** amounts to:
(21)

$$P = (1 - \frac{dw}{\Delta w})^N$$

With **N** being finite, **DELTA w** not shrinking to zero, and with **dw** shrinking to zero, **P** approaches the value of **1** . In other words: if **dw** is allowed to shrink towards zero, the chances for an interval **dw** to host just a single molecule approach zero.

It has to be conceded that the error, though of first order, is very small (due to the very large number of molecules). Boltzmann nevertheless feels a need for replacing **dwdz** by another differential: When drawing a two-dimensional system of coordinates with **z** and **w** as the two perpendicular axes, one can recognize that the surface area **F** in which the molecules described are located is identical in size (not in shape) whatever point in time is depicted. In an effort to take advantage of this constancy, Boltzmann modifies equation (15) by replacing the product **dw dz** with the area **F** formed in the coordinate system (**w,z**). By means of this artifice the widths of the intervals remain unchanged.

Boltzmann continues his train of thoughts as follows: As he has proved the differentials **dudvdF** to be constant over time despite the influence of gravity, the remaining thing to do is to show that the term
(22)

$$C\,e^{\wedge}[-h(u^2+v^2+w^2)-kz]$$

equals the term
(23)

$$C\,e^{\wedge}\{-h[u^2+v^2+(w+gt)^2]-k(z-wt-gt^2/2)\}$$

(the former term describes the state of the gas at time zero, the latter term describes the state of the gas at a later point in time, after **z** and **w** have changed as a result of gravitation). If this task could be coped with, Boltzmann would have proved that equation (15) is able to describe the state of the gas even under the influence of gravity.

Boltzmann shows that (22) and (23) are eqivalent as soon as the constant **k** is substituted by **2gh**. However, he has to be criticized for not having proved that his substitution **k = 2gh**, beside from being sucessful in a mathematical sense, is consistent with other laws of physics. In other words: if **k=2gh** contradicted with other laws of physics, that substitution could not be useful in physics.

Whether or not the equation **k = 2gh** is empirically correct, can only be decided after the constants **C, h** and **k** appearing in equation (15) are determined (without using the equation **k=2gh**). This is what we will do in the following.

Boltzmann's first starting point is his formula of the distribution of velocities, which is

(24)

$$dN(u,v,w) = N\left(\frac{m_t}{2\pi k_B T}\right)^{3/2} e^{-\frac{m_t}{2k_B T}(u^2+v^2+w^2)} \, du \, dv \, dw$$

with k_B denoting the Boltzmann constant, and m_t denoting the mass of a molecule. Because of $1/2 \, m\_t \, v\_tot\ ^2$(overlined) $= 3/2 \, k\_B \, T$, (24) can be turned into

(25)

$$(dN)_1 = N\left(\frac{3}{2\pi v^2_{tot}}\right)^{3/2} e^{-\frac{3}{2v^2_{tot}}(u^2+v^2+w^2)} \, du \, dv \, dw$$

where **v_tot² (overlined)** denotes the mean square of the molecules' total velocities.

Now we turn to the second starting point of Boltzmann's formula. For an isothermal gas subject to gravity, the pressure **p** is a function of the height **z** above the ground:

(26)

$$p = p_0 \, e^{-\frac{q_0}{p_0} g z}$$

The term q_0 denotes the density of the gas at the base, p_0 denotes the pressure of the gas at the base, g denotes the gravitational acceleration. For a gas of constant temperature, the number of molecules is proportional to the pressure of the gas, provided the volume of the gas, that is the height dz of the stratum considered, remains constant. On the other hand, the number of molecules is proportional to the volume of the gas, that is to the height dz of the stratum multiplied by the area of the base of the column, provided the pressure remains constant. As a consequence, the number of molecules is proportional to the product of pressure and volume. Hence, for a gas of constant temperature, we have (27)

$$(dN)_2 = a\, p\, A\, dz$$

with "a" denoting the (temperature dependent) factor of proportionality, equivalent to the ratio of the number particles per unit volume and pressure. The term "A" denotes the area of the base of the column. When, in equation (27), **p** is substituted by the right side of equation (26), and if, in addition, both sides are multiplied by **1/N**, we arrive at

(28)

$$\frac{(dN)_2}{N} = \frac{1}{N}\, a\, p_0\, e^{-\frac{q_0}{p_0} g\, z}\, A\, dz$$

Searching for the number of molecules the velocities of which range within u and u + du, v and v+ dv, w and w+ dw , and the z- coordinates of which range between z and z + dz , the right sides of the two equations (25) and (28) must be multiplied by each other. So we get (29)

$$dN = a\, p_0\, A\, \left(\frac{3}{2\pi \bar{v}^2_{tot}}\right)^{3/2} e^{-\frac{3}{2\bar{v}^2_{tot}}(u^2+v^2+w^2) - \frac{q_0}{p_0} g\, z}\, du\, dv\, dw\, dz$$

As a consequence, the constants **C, h** and **k** can be determined as follows:

(30)

$$C = a\, p_0\, A\, \left(\frac{3}{2\pi \bar{v}^2_{tot}}\right)^{3/2}$$

(31)

$$h = \frac{3}{2\overline{v}^2_{tot}}$$

(32)

$$k = \frac{q_0}{p_0} g$$

The constant **k** must not be confused with the general Boltzmann constant **k_B** or with the constant **k** appearing in Poisson's equation.) We realize that **C** and **h** contain **v_ tot2** (**overlined**) as a factor, which is an expression of the temperature of the gas. Moreover, **C** contains **T** explicitly, since the norming constant of the barometric formula contained **T**.

We are now able to decide whether or not Boltzmann's substitution **k = 2gh** is empirically justified. When, in **k = 2gh**, the terms **k** and **h** are substituted according to (31) and (32), we get:

(33)

$$k = 2gh \quad - \quad \frac{q_0}{p_0} g = 2g \frac{3}{2\overline{v}^2_{tot}}$$

In the following, the density q_0 is expressed as M/V_0 ; V_ 0 is the specific volume of 1 mol of the gas at the bottom, M denotes the mass (in kg) of 1 mol; N_ L denotes Loschmidt's number, m_t [kg] denotes the mass of a single molecule, R denotes the gas constant, T the temperature, k_B denotes Boltzmann's general constant. So (33) turns into

(34)

$$\frac{M}{p_0 V_0} = \frac{m_t}{k_B T}$$

or

(35)

$$p_0 V_0 = \frac{M \, k_B \, T}{m_t} = N_L \, k_B \, T = RT$$

{Unit of $p_0 V_0$ = J/mol ; unit of $(M \, k_B \, T)/m_t$ = [(kg/mol)(J/K) K]/kg = J/mol}

Thus, when reversing the order of equations, Boltzmann's substitution **k = 2gh** can be derived from the basic equation of the ideal gas.

Though the outcome of our calculations appears to approve Boltzmann's reasoning, an evident weakness in his reasoning is revealed when taking a look at the constants **h** and **C**. Both are determined by the average square of the all molecules' total velocity v_{tot}, that is by the average kinetic energy of the whole gas. That square of velocity, however, is affected by the change of the velocity **w** of the molecules considered, which, in turn, is a result of gravity. Boltzmann fails to prove that a change in the average square of all molecules' velocities (caused by the change in vertical velocity of the molecules considered) is avoided by the compensating behaviour of other groups of molecules. The only "candidates" who appear to be at hand for such a compensation are the molecules the velocities of which range from **u** to **u+du,** and from **v** to **v+dv** , and from **-w** to **-w-dw** (ascending molecules). Though, in a given interval **dt**, their vertical velocity is reduced by the same amount by which the vertical velocity of the molecules which are considered by Boltzmann (descending molecules) is increased, the loss in kinetic energy of the ascending group, that is

$$1/2 \, m[\, w_0{}^2 - (w_0 - \text{Delta } w)^2]$$

for an individual molecule in the ascending group, falls short of the increase in kinetic energy of the descending group, that is

$$1/2 \, m[(w_0 + \text{Delta } w)^2 - w_0{}^2]$$

for an individual molecule in the descending group. Hence, Boltzmann's reasoning is incomplete. He defines **C** and **h** as constants though they could be variables. The incompleteness vanishes only when the gravity **g** approaches zero, that is when v^2_{tot} **(overlined)** remains almost invariant.

Moreover, Boltzmann (like Walton and Garrod) considers a one-dimensional gas only. He dismisses the possibility that the molecules may gain height at the expense of all three components of velocity.

Loschmidt summarized his own critique on Boltzmann's reflections in the following way:

"First. The results can, because of their starting point, only claim an approximate validity for the real gases, and thus have to be considered worthless for the foundation of strictly valid statements in the mechanical theory of heat. Second. The equality of temperature in all strata of a mass of gas is by no means warranted by the formula found, since for such a warranty it is, in addition, required to assume that the molecules have identical limits of the components of velocity throughout the whole mass of gas. Such an assumption, however, is not admissible with regard to systems of few molecules, and is not proved with regard to those of many ones." [8, 9]

Boltzmann's second attempt

Some twenty years after his discussion with Loschmidt, Boltzmann revised his argumentation in his Lectures on Gas Theory (1896-98). Seemingly, he realized that his first proof was not valid, as he openly admitted that the alleged constants might be variables [10] , just as we did in the foregoing chapter. Of course, he tries to give a new proof of the thesis that these constants do not depend on height. Boltzmann's second attempt to prove the uniformity of temperature shall be discussed in the following.

In Chapter 2 of the Lectures on Gas Theory he writes (some Greek letters have been replaced by Latin ones): [11]

"We fix our attention on the parallelepiped representing all space points whose coordinates lie between

(97)

$$x \text{ and } x + dx, y \text{ and } y + dy, z \text{ and } z + dz.$$

... The velocity of each m-molecule that finds itself at time t in this parallelepiped shall be represented by a line starting at the origin, and the other end point C of this line will again be called the velocity point of the molecule. Its rectangular coordinates are equal to the components u, v, w of the velocity of the molecule in those coordinate directions.

We now construct a second rectangular parallelepiped, which includes all points whose coordinates lie between the limits

(98)

$$u \text{ and } u + du, v \text{ and } v + dv, w \text{ and } w + dw.$$

... The m-molecules that are in [the first parallelepiped] dxdydz at time t and whose velocity points lie in [the second parallelepiped] dudvdw at the same time will again be called the specified molecules, or the dn molecules. Their number is clearly proportional to the product dxdydz dudvdw. Then all volume elements immediately adjacent to dxdydz find themselves subject to similar conditions, so that in a parallelepiped twice as large there will be twice as many molecules. We can therefore set this number equal to

(99)

$$dn = f(x,y,z,u,v,w,t) \; dxdydzdudvdw.$$

... We shall now allow a very short time dt to elapse, and during this time we keep the size and position of dxdydz and dudvdw completely unchanged. The number of molecules that satisfy the condition (97) and (98) at time t + dt is, according to Equation (99),

$$dn' = f(x,y,z,u,v,w,t+dt) \; dxdydzdudvdw$$

and the total increase experienced by dn during time dt is

$$dn' - dn = (\partial f/\partial t) \; dxdydzdudvdwdt."$$

Now Boltzmann considers the molecules that are located left from the parallelepiped dxdxdz in real space. Provided they have an exact velocity u (in fact their velocity ranges from u to u + du) all molecules within a distance of udt will manage to fly through the left face of our parallelepiped within that temporal interval dt. Their number equals (36)

f(x,y,z,u,v,w,t) udtdydzdudvdw

The function **f** is an expression of the particle density at the depicted spot in the combined, six-dimensional space; or, in other words: an expression of the density of particles with a certain velocity at the depicted spot in the three dimensional real space.

Similarly, the number of molecules leaving the parallelepiped **dxdydz** through its right face will be

(37)

$$f(x + dx,y,z,u,v,w,t) \; udtdydzdudvdw$$

The difference between the two expressions amounts to

(38)

$$(\partial f/\partial x)dx \; udtdydzdudvdw$$

with $(\partial f/\partial x)dx$ denoting the partial differential of **f**.

By similar arguments for the four sides of the parallelepiped one finds that during time **dt**

(39)

$$- (u \; \partial f/\partial x + v \; \partial f/\partial y + w \; \partial f/\partial z) \; dxdydzdudvdwdt$$

more molecules enter the parallelepiped dxdydz than leave it as a consequence of their velocities.

So far, the change in speed of the molecules during dt has not been taken into account. Boltzmann tries to do this by showing (as a first step) that in velocity space, the points representing molecules move with constant speed when subject to acceleration, and are at rest when not subject to acceleration. So he employs completely similar arguments with respect to the motion of velocity points (through the parallelepiped **dudvdw**) as he previously did for the motion of the molecules themselves in real space (through **dxdydz**). In the former case, the velocity of a molecule was assumed to be constant over the path of the molecules during **dt** (the function **f** was differentiated after **x** only), now the spatial density (for any given class of molecules having speeds between **u** and **u+du**) is assumed to be constant over the path of the molecules during **dt** (the function **f** is differentiated after **u** only).

So Boltzmann finds that in all

(40)

$$- (X \; \partial f/\partial u + Y \; \partial f/\partial v + Z \; \partial f/\partial w) \; dxdydzdudvdwdt$$

more velocity points enter dudvdw than leave it as a consequence of an acceleration the molecules are subject to. **X,Y** and **Z** are the components of that acceleration.

As a next step, Boltzmann argues that for a system to be in an equilibrium, the sum of both expressions has to be zero, since the number of molecules in the six-dimensional parallelepiped has to be constant over time.

But admitted by Boltzmann himself, his expressions contain an inaccuracy which has already been mentioned above. The inaccuracy shall be demonstrated: Imagine a box (of differential size) having a hole in the ceiling and in its bottom. Particles are dropped from above (in constant temporal intervals), entering the box through the hole in the ceiling and leaving it through the second hole. All molecules share the same direction of motion (**+z**). Though the temporal gap between those molecules that have the same speed stays constant (when passing the hole in the bottom of the box), their mutual spatial distance increases while they are falling and gathering speed. With the spatial particle density thus decreasing towards the bottom of the box (= defined as the positive direction of **z**) for a given speed, **ðf/ðz** is different from zero (negative). Boltzmann's formula, which (because of the fact that the particle density does not depend on **x,y,u**, or **v**) reduces to (41)

$$- w\, ðf/ðz\, dzdwdt$$

would thus yield a net increase in the number of molecules inside the box for all velocities (in contrast with reality).

Let us see whether or not the inaccuracy will disappear after an integration (as was asserted by Boltzmann). An integration over w (zero to infinity) would lead to the expression:

(42)

$$a(z)\, dzdt$$

To elucidate: ðf/ðz may depend on w and on z. This is why an integration over w results in a function of z. The value of that function a(z) can easily be determined for the ceiling of the box (which shall now be assumed to be no longer of differential size only). At this particular location, the term "- w ðf/ðz dw" is positive for all positive values of w (since the density of particles is decreasing with z whatever group of molecules is considered). Hence a(z) , which is the sum of all these numbers, has to be positive for this particular value of z. The same applies to all locations z between the ceiling and the bottom of the box. So an integration over z leads to

(43)

$$C\, dt$$

with the constant **C** being greater than zero.

After a final integration over **t**, we would still have a positive number different from zero.

Let us now turn to the second term which is supposed to yield the number of molecules entering or leaving **dudvdw** (and hence the sixth-dimensional parallelepiped) as a consequence of acceleration. Again, there is an inaccuracy contained in that expression (admitted by Boltzmann himself), and we will see whether or not it can be neglected (as asserted by Boltzmann) after an integration (or whether it is even capable of compensating the the inaccuracy of the first expression). We will therefore return to our box. The expression reduces to (44)

$$- Z \, \eth f/\eth w \; dzdwdt$$

An integration of (44) over w would yield
(45)

$$b(z) \, dzdt$$

since **ðf/ðw** may depend on **z** and on **w**. As we can arrange the distribution function **f** according to our wishes, we determine that there are fewer molecules with higher speeds than with lower speeds; also we determine that **ðf/ðw** shall be negative for all positive values of **w** (no matter which spot within the box is considered). Hence, at a particular location **z** between the ceiling and the bottom, the expression "- **Z ðf/ðw dw**" is a positive number for any positive velocity w. Then **b(z)** must, at this location **z** , be a positive number, since it is the sum of all these previous numbers. The same applies to all locations **z** between the bottom and the ceiling. This is why an integration over **z** yields (46)

$$K \, dt$$

with **K** representing a positive constant. A final integration over **t** would again lead to a positive numerical value.

As a result, the sum of Boltzmann's differential expressions yield a result different from zero though the system we constructed is in a state of equilibrium. Therefore the method he suggests for describing a state of equilibrium is not convincing. This is just one reason why his second attempt, which is based on that method, fails.

There is, of course, a second reason: Boltzmann does not realize that it is the mutual collisions of the molecules that is responsible for the stratification of temperature in a vertical column of gas subject to gravity.

A late completion of Boltzmann's homage paid to Loschmidt; consequences for the nature of time

When Josef Loschmidt died in 1895, Boltzmann held a memorial speech addressed to the Chemical-Physical Society of Vienna on the 29th of October, 1895. On this occasion, he rated the computation of the number of molecules contained in a unit volume to be Loschmidt's greatest discovery [12] . Such a rating must be contradicted. A discovery at least equivalent to the one mentioned by Boltzmann is the compatibility of the second law of thermodynamics and the perpetual motion machine of the second kind. If Loschmidt's discovery of that compatibility had been widely accepted in those days, the evolution of energy technology might have been a different one. Unfortunately, Loschmidt's arguments in favor of the stratification of temperature in a gas subject to gravity do not provide a strict proof. With a strict proof at hand, he could have spread his thesis with a greater psychological effect.

Furthermore, Loschmidt pointed out that the second law could be derived from the principle of least action. To put it differently: he replaced the original foundation of the second law (that is the axiom of the impossibility of a perpetual motion machine of the second kind) by a different one. Doing so he referred to Boltzmann, who had already displayed such foundation in his article *"Über die mechanische Bedeutung des 2. Hauptsatzes der Wärmetheorie."* It can be left undecided whether or not the derivation of the second law from the principle of least action is strictly convincing. In a recent article, G. Bierhalter, who has published several articles on the history of the second law, doubted the strictness of such reasoning. [13] . In any case, the second law and a perpetual motion machine of the second kind are compatible, as soon as we no longer define the second law as it has been usual. In that common form the second law asserts that for a quantity of heat to pass from one body to another, the temperature of the receiving body must be lower than that of the donating body. Rather, one should formulate (as already proposed above) the second law as the assertion that heat does not pass from one body to another if such a passage would destroy a state of equilibrium. A state of equilibrium, in turn, is a state in which the net flow of heat has become zero. To put it differently: If the flow of heat between two bodies has come to a standstill, that flow will not start again by itself. In most cases, the two bodies will then have the same temperature; in some cases, however, the temperatures of the two bodies will be different from each other. The concept of entropy should be confined to cases in which real heat flows are taking place. Of course, this will deprive entropy of its character as a variable of state, and the Second Law could no longer be expressed by applying the notion of entropy.

Of course, defining the Second Law the way just proposed is almost identical with the defintion of the word "equilibrium." Nevertheless, the consequences of an adoption of that definition are enormous: The fact that any difference in density and temperature within a quantity of gas will vanish after a while, has to be ascribed to the *original state of matter*, not to the Second Law. In that respect, there is a resemblance between temperature and gravity: The fact that water in a cup, when stirred, will climb the walls of the cup as a result of the centrifugal "force" at work, is — as General Relativity tells us — a result of the special way the distant stellar masses of the universe are distributed. In much the same way, the fact that cold water, when added to a cup containing hot coffee, will mix with the coffee to form a liquid of uniform temperature, is a result of the state of matter in the universe billions of years ago. The often discussed paradox, that is the question why temperature differences within an ideal gas will always vanish though all motions of the molecules are reversible (so that increases in temperature differences should be as frequent as reductions of these differences) is thereby resolved: The initial state of the gas to start with (which itself is causally dependent on prior states of things) isn't of the right kind for generating temperature differences.

Moreover, this recognition gives rise to revisit Boltzmann's famous dispute with Zermelo. In a universe endless both in time and in space, he argues, there must exist "islands" in which, by random processes, matter is organized, whereas the universe is barren and at uniform temperature elsewhere. Living beings (including intelligent machines) on such an island will *define* the arrow of time by saying that the future is the *less* organized state, while the past is the *more* organized state of their island (there is no physical definition of the arrow of time other than this one, since the laws of physics are time-symmetric). The universe is hence in posession of different states, but is lacking of an intrinsic ordering of these states by the category earlier / later. Instead, such an ordering is extrinsic. It seems that Boltzmann's view of the arrow of time is quite correct despite the fact that cosmology, by assuming the Big-Bang at the "origin" of the universe, is dismissing the assumption of a universe endless both in space and time.

This view might even shed some light on the evolution of life. Given there is no intrinsic tendency of nature of degrading its state of order, an increase in order within parts of the physical world is something quite "natural" that does not need any special mechanism to account for. In as much as man — as part of nature — could, by use of a multitude of perpetual motion machines of the second kind, increase the order of things surrounding him, so could nature by other means.

Finally, for reasons of clarity, the following reflection should be underlined: The dismissal of an intrinsic ordering of earlier / later follows from the *statististical concept* of the Second Law already. In other words: For such a view to be adopted, it is not necessary to allow the construction of a perpetual motion machine of the second kind. One could as well assert that the "initial" conditions of matter were such that the operation of a perpetual motion machine is excluded for all times, or at least for eons of time, as the causal chains, given those initial conditions, can never lead to such a phenomenon. The proof of the extrinsic character of the ordering earlier / later could thus be given by the argument that the reversal paradox (mentioned above) cannot be resolved otherwise. The benefit of a perpetual motion machine of the second kind, in this context, lies in the fact that it makes the extrinsic character of an ordering earlier / later more evident, since the existence of such a machine is equivalent to saying that nature is *not* governed by a law that — *regardless* of the initial conditions to start from — will lead to a degradation of order, but is governed by time-symmetrical laws only that do not allow to define an intrinsic arrow of time.

It is hardly known that Ernst Mach, too, was skeptical as regards the reach of the Second Law. He objected to the generalization of the original concept of entropy, that is the amount of heat received or given off by a body divided by its temperature, and was skeptical that an increase in entropy (in a sense just described) could be paralleled with an increase in disordered motions of particles. In Chapter 102 of his "Principles of the Theory of Heat," he wrote (my own translation from German): "*The mechanical view of the Second Law, which distinguishes ordered and disordered motions by paralleling the increase in entropy with the increase in disordered motions at the expense of ordered motions, appears to be quite artificial. Taking into account that a real analogue of the increase in entropy does not exist in a purely mechanical system made up of perfectly elastic atoms, one can hardly reject the idea that an infringement of the Second Law should be quite possible — even without any help from demons — given such a mechanical system were indeed the basis of the heat phenomena.*"

Loschmidt had the following vision for the future: "*Thereby the terroristic nimbus of the second law is destroyed, a nimbus which makes that second law appear as the annihilating principle of all life in the universe, and at the same time we are confronted with the comforting perspective that, as far as the conversion of heat into work is concerned, mankind will not solely be dependent on the intervention of coal or of the sun, but will have available an inexhaustable resource of convertible heat at all times.*" [15]

NOTES

1) Claude Garrod: Statistical mechanics and thermodynamics, Oxford University Press, 1995

1a) See also: R. Baierlein, "How Entropy got its Name," *American Journal of Physics*, 60, 1151.

2) See E. Pertigen: Der Teufel in der Physik - Eine Kulturgeschichte des Perpetuum Mobile, (Berlin: Verlag für Reisen und Wissenschaft 1988).

3) Transactions of the Royal Society of Edinburgh 20 (1851), 265.

4) J. Loschmidt, "Über den Zustand des Wärmegleichgewichts eines Systems von Körpern mit Rücksicht auf die Schwerkraft I," Sitzungsberichte der mathematisch - naturwissenschaftlichen Classe der Kaiserlichen Akademie der Wissenschaften zu Wien 73.2 (1876), 135.

5) J.C. Maxwell, "On the Dynamical Theory of Gases," *The London, Edinburgh, and Dublin Philosophical Magazine* and *Journal of Science 35* (1868), 215/216. Some years later, Maxwell repeated his argument. In his book *Theory of Heat*, published in London in 1877, he writes (p. 320): "...if two vertical columns of different substances stand on the same perfectly conducting horizontal plate, the temperature of the bottom of each column will be the same; and if each column is in thermal equilibrium of itself, the temperatures at all equal heights must be the same. In fact, if the temperatures of the tops of the two columns were different, we might drive an engine with this difference of temperature, and the refuse heat would pass down the colder column, through the conducting plate, and up the warmer column; and this would go on till all the heat was converted into work, contrary to the second law of thermodynamics. But we know that if one of the columns is gaseous, its temperature is uniform. Hence that of the other must be uniform, whatever its material." Thus Maxwell did not modify his assertion that if there were a temperature gradation in a column of gas subject to gravity, a perpetual motion machine of the second kind would become possible.

6) A.J. Walton, "Archimedes' Principle in Gases," in: *Contemp. Phys.*, 1969, Vol. 10, No. 2

6) Loschmidt, "Über den Zustand des Wärmegleichgewichts...I," p. 133.

7) See L. Boltzmann, "Über die Aufstellung und Integration von Gleichungen, welche die Molekularbewegung von Gasen bestimmen" in L. Boltzmann, Wissenschaftliche Abhandlungen, edited by F. Hasenöhrl, vol. 2 (Leipzig: Barth 1909), p. 56ff.

8) J. Loschmidt, "Über den Zustand des Wärmegleichgewichts eines Systems von Körpern mit Rücksicht auf die Schwerkraft IV,"

Sitzungsberichte der Kaiserlichen Akademie der Wissenschaften zu Wien 76.2 (1877), 225.

9) An argument similar to Boltzmann's can be found with S.H. Burbury, "Equilibrium of Temperature in a Vertical Column of Gas," *Nature*, Vol. 12 -1875-, p. 107.

10) L. Boltzmann: Lectures on Gas Theory, Dover Publ. 1964, par. 19, p.141

11) L. Boltzmann: Lectures on Gas Theory, Dover Publ., par. 15

12) See L. Boltzmann, "Zur Erinnerung an Josef Loschmidt," in L. Boltzmann, Populäre Schriften (Leipzig: Barth 1905).

13) G. Bierhalter, "Von L. Boltzmann bis J.J. Thomson: die Versuche einer mechanischen Grundlegung der Thermodynamik," Archive for the History of Exact Science 44 (1992), 25-72.

14) See Loschmidt, "Über den Zustand des Wärmegleichgewichts... I," p. 141.

15) Loschmidt, "Über den Zustand des Wärmegleichgewichts... I," p. 135.

Acknowledgments

There were several plagiarism bru-ha-has going around about the time I was contemplating how I should write these few pages of acknowledgements and who should be included. It became obvious to me as I was listening to all this fuss of who stole what prose from whom, that this book really was a team effort and I'll surely forget someone who really has been instrumental in one of these novel ideas, or encouraged me to get off my duff to actually churn this out. So a pen name seemed an appropriate way to acknowledge all those who have helped me in so many ways. I sincerely express my apologies to those I neglect to acknowledge.

This book is about encouraging new generations to explore. So I'll start with those who encouraged me to not shy away from exploring tabooed science. Peter Cudmore, Edward Teller and Barry Boehm all went out of their way to encourage me to explore novel ideas and forget the advice of naysayers. Peter Cudmore took me into his house as an AFS exchange student. Many days he would drive three young hellions – cleverly disguised in immaculate three-piece suits – to the prestigious Camberwell Grammar school for boys located in Melbourne, Australia. I remember clearly the few times he would look into his Jaguar's mirror and ask, "Drake do you want to go to school today, or do you want to learn something really exciting?" The rest of the car would pipe up how they wanted the excitement option too while Peter shoved them out - and I listened to their cursing the unfairness of it all. Then it was off to who knows where as he hammered on and on about the evil of naysayers...how he made his fortune mining molybdenum ore...and thank god for that rotting son of a bitch Hitler, whose only redeeming quality was that he changed Peter's otherwise worthless molybdenum ore into something more valuable than gold for the British Empire!

But it was his constant theme of avoiding those filled with envy, and following your optimistic instincts that I remember only second to his friendly smile. This constant drumbeat about the dangers of envy was from a man whose cunning was second to none.

Now I'll fast-forward to UCLA, which was a fantastic and frustrating time for me. Edward Teller told one of my professors to summon me to his office soon after I had presented my paper on the hydrogen-car project. My paper's theme was clearly "outside the box" of the project's agenda. Teller greeted me with a broad smile from behind a classic wooden desk as I settled into an uncomfortable wooden chair that reminded me of Methodist pews designed to keep the flock from dozing off. It's impossible for me to exaggerate Teller's optimistic lecture – and make no mistake, it was a lecture. He smothered me with praise and warnings. His central theme was the dangers of envy and naysayers as he interjected comments and questions about my torque equations describing combinations of different fuel mixtures.

I sat in awe that a man his age could have such enthusiasm for life and anger for envy. I was also amazed that he was oblivious to the long strands of grey hairs coming out of his ears and nostrils. I smiled when I thought, "Man – this guy is really smoking." He really gathered a full head of steam as he ranted on about universities having way too many tenured who harbor nothing in their hearts but petty envy for the likes of us. I was wondering if his anger came from knowing how I got my noteworthy F in math analysis, which in turn reminded him of many past battles fought with less talented physicists like Robert Oppenheimer - whose prestige came from presidential favoritism, birthright and social politics rather than from Oppenheimer's mediocre skills in physics. Regardless, this unique elder lavishing encouragement on me will never be forgotten. As I was departing, I recall thinking how Edward Teller's and Peter Cudmore's love for optimism and disdain for envy were remarkably similar...not to mention their strong desire to pass these thoughts on to the next generation.

Barry Boehm's encouragement was so low-keyed that it stands in sharp contrast to the thunder of Cudmore and Teller. I must admit that the importance of these three men cannot be over-stated. I would have likely shied away from such a speculative book without their encouragement to take chances and try a little trailblazing. As an undergrad, I'll never forget Barry giving me a tour of Rand while he was head of its computer/math department. He was quick to point out the many employment virtues of Rand...like...its close proximity to beach volleyball courts, surf and bikinis. But my favorite memory was when he and a colleague gleefully ran off a laundry list of Rand's bloopers...from predicting the first heart transplant....to predicting when there will likely be an oversupply of super computers in the world. Well, so much for the "best and brightest" *always* getting it right. I enjoy his unique humor. His unpretentious nature was a refreshing reprieve from the elite arrogance often found on campuses. Barry has always encouraged me to tackle uncertainty head-on and I really appreciate his review of these speculative theories. His suggestions have truly improved this book. I also wish to thank Daniel Sheehan who gave me more typical suggestions on the physics aspects of this book.

On the medical front, Steve Pandol offered great medical insight as we collaborated on red wine research together. His tale of political intrigue as

Director of Research and Development for the VA section of the UCLA/VA medical facility was truly eye opening and I am grateful to him for letting me witness, and also collaborate on medical research with him firsthand. The same is also true for David Klurfeld, Kosta Arger, Janine Krivokapich, Mohamad Navab, David Fitzpatrick and Anna Gukovskaya, who also helped me as collaborators on red wine medical research. Bo Saxberg was also kind enough to offer great insight about the medical world. His breadth of wisdom is noteworthy. Greg Misbach, John Frazee and George Teitelbaum were kind enough to give me a close perspective of the medical/biotech world from a surgeon's perspective. There are many other doctors who were happy to help me with their insight into the medical world. I thank them all.

Nora De Caprio is an amalgamate of Susan Stern, Peter Aleshire, Chalin Smith and Erin De Caprio. Sorry about feminizing you, Pete, but I secretly suspect that there is a beautiful gown way in the back of your closet that just might fit you surprisingly well (just kidding, its Nora just because majority rules). They all can run circles around my technical writing skills and have contributed greatly in taming the ADHD herein. Others who I would like to thank for comments and suggestions are Laurie Lappin, Carrie Radeleff, Kevin Boylan and Craig Smith.

I am a product of my education and would feel remiss if I failed to mention some great teachers of mine. Palm Springs had many remarkable teachers in the 60s-70s, like Francis Neunan, Hal Bentley, Max Morris, Muriel Danforth, Betty Rosine, Doris Gardner, Charles O'Dahl, Letha Cote, James Vilmann, Lloyd Saatijan and Cecil Jones. I was also fortunate to attend at UCLA noteworthy math lectures presented by James White, Theodore Gamelin, Diane Schwartz and Larry Miller. There is no way this book would have been written without their dedication to educate my generation...and while we're on the subject of my education, a central theme of this book is found in earlier works of Thomas Kuhn, James Burke and Thomas Peters.[47]

Finally, and most importantly, is my immediate family: Mom and Dad, who can go toe-to-toe with Cudmore and Teller in encouraging exploration, and my loving wife, Pamela, and daughters, Kerry and Annika, who were kind enough to illustrate, edit and occasionally ask, "Why did you choose to grow grapes and flowers instead of pursuing a career in mathematical physics or dammed lies?" I thank you one and all.

Drake Larson

---

[47] "The Structure of Scientific Revolutions," "The Day the Universe Changed," and "In Search of Excellence," respectively.

Index